Atmospheric Aerosol
Source/Air Quality Relationships

Edward S. Macias, EDITOR
Washington University

Philip K. Hopke, EDITOR
University of Illinois

Based on a symposium jointly

sponsored by the Divisions of

Nuclear Chemistry and Technology

and Environmental Chemistry

at the Second Chemical Congress

of the North American Continent

(180th ACS National Meeting),

Las Vegas, Nevada,

August 27–29, 1980.

A C S S Y M P O S I U M S E R I E S **167**

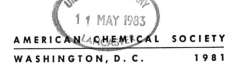

AMERICAN CHEMICAL SOCIETY
WASHINGTON, D. C.   1981

Library of Congress CIP Data
Atmospheric aerosol.
(ACS symposium series, ISSN 0097-6156; 167)

"Based on a symposium jointly sponsored by the
Divisions of Nuclear Chemistry and Technology and
Environmental Chemistry at the Second Chemical Con-
gress of the North American Continent (180th ACS
National Meeting),Las Vegas, Nevada, August 27–29,
1980."
Includes bibliographies and index.

1. Aerosols—Addresses, essays, lectures. 2. Air quality—Mathematical models—Addresses, essays, lectures.
3. Air quality—United States—Addresses, essays, lectures. 4. Atmospheric chemistry—Addresses, essays,
lectures.
I. Macias, Edward S., 1944- . II. Hopke, Philip
K., 1944- . III. American Chemical Society. Division of Nuclear Chemistry and Technology. IV. American Chemical Society. Division of Environmental
Chemistry. V. Chemical Congress of the North American Continent (2nd: 1980: Las Vegas, Nev.) VI.
Series.

QC882.A868   628.5'3   81-10960
ISBN 0-8412-0646-5 AACR2 ASCMC 8 167 1–359
                                          1981

ACS Symposium Series

M. Joan Comstock, *Series Editor*

FOREWORD

The ACS SYMPOSIUM SERIES was founded in 1974 to provide
a medium for publishing symposia quickly in book form. The
format of the Series parallels that of the continuing ADVANCES
IN CHEMISTRY SERIES except that in order to save time the
papers are not typeset but are reproduced as they are sub-
mitted by the authors in camera-ready form. Papers are reviewed under the supervision of the Editors with the assistance
of the Series Advisory Board and are selected to maintain the
integrity of the symposia; however, verbatim reproductions of
previously published papers are not accepted. Both reviews
and reports of research are acceptable since symposia may
embrace both types of presentation.

CONTENTS

PREFACE

One of the central problems in air pollution research and control is to determine the quantitative relationship between ambient air quality and emission of pollutants from sources. Effective strategies to control pollutants can not be devised without this information. This question has been mainly addressed in the past with source-oriented techniques such as emission inventories and predictive diffusion models with which one traces pollutants from source to receptor. More recently, much effort has been directed toward developing receptor-oriented models that start with the receptor and reconstruct the source contributions. As is the case with much of air pollutant research, improvements in pollutant chemical analysis techniques have greatly enhanced the results of receptor modeling.

Here we have restricted our attention to atmospheric aerosols (particulate matter) because of the crucial role these particles play in adverse health effects, visibility reduction, soiling, and acid rain—the most serious effects of air pollution. However, it should be noted that many of the techniques discussed in this book also can be applied to gas-phase species.

The papers in this volume were presented at the ACS symposium "Chemical Composition of Atmospheric Aerosol: Source/Air Quality Relationships" sponsored by the Divisions of Nuclear Chemistry and Technology and Environmental Chemistry. This combination reflects the interdisciplinary nature of much of this work which draws from such diverse fields as nuclear chemistry, chemical and mechanical engineering, environmental science and engineering, and applied math.

The symposium was divided into four subject areas, and this volume follows that general format. The first group of chapters reviews and describes many of the recent modeling efforts. The next section is devoted to source characterization studies, while the third group includes chapters concerned with carbonaceous aerosols—both source apportionment and measurement techniques. The final section describes the results of several field studies in areas of the United States and China where wind-blown dust is a serious problem.

These chapters provide only a sampling of activity in this area. However, several of the chapters review the extensive literature in this field. The chapter by Watson et al. reports the state of the art of receptor modeling as determined by the consensus of the participants at a recent workshop on the subject. Therefore, we hope that this book will serve not only as a record of the work presented at the ACS symposium but also as a general guide to the field as of fall 1980.

We want to thank all of the participants in the two and a half day symposium. A total of 28 papers were presented at that time of which 18 are included here. It is our hope that other symposia on various aspects of this very lively field will be held in the future.

EDWARD S. MACIAS
Washington University
St. Louis, MO 63130

PHILIP K. HOPKE
University of Illinois
Urbana, IL 61801

May 1, 1981

New Developments in Receptor Modeling Theory

S. K. FRIEDLANDER

Department of Chemical, Nuclear, and Thermal Engineering,
University of California—Los Angeles, Los Angeles, CA 90024

Receptor modeling has become an important tool for
developing particulate air pollution control
strategies. Current receptor models for ambient
aerosol source resolution begin with the measured
chemical properties of the aerosols at a given site
and infer the mass contributions of various sources
to the total measured mass. Rate processes are not
involved in these models. In this paper, the theory
is extended to the resolution of the visibility
degrading components of the aerosol and to
chemically reactive families of chemical compounds.
Conditions are discussed under which a linear
relationship with constant coefficients exists
between the aerosol light extinction coefficient and
source mass contributions. Linear models for
reactive chemical species are set-up and applied to
polycyclic aromatic hydrocarbons using data from the
literature. In both cases-light extinction and
chemical reactivity-it is useful to introduce rate
process models in developing the theory.

Emission inventories are often used to develop control
strategies for particulate pollution, but there are difficulties

with their use including secondary aerosol formation, long range
transport and particle deposition from the atmosphere according to
size. There are, in addition, important hard-to-characterize
sources such as soil dust and the marine aerosol.

A different approach which also starts from the
characteristics of the emissions is able to deal with some of
these difficulties. Aerosol properties can be described by means
of distribution functions with respect to particle size and
chemical composition. The distribution functions change with time
and space as a result of various atmospheric processes, and the
dynamics of the aerosol can be described mathematically by certain
equations which take into account particle growth, coagulation and
sedimentation (1, Chap. 10). These equations can be solved if the
wind field, particle deposition velocity and rates of
gas-to-particle conversion are known, to predict the properties of
the aerosol downwind from emission sources. This approach is
known as dispersion modeling.

While such calculations can be carried out in principle, they
are in fact rarely possible in the detail needed for developing
reliable air quality/emission source relationships for particulate
pollution. Dispersion modeling however, is necessary to predict
the air quality effects of a new source which is to be located in
a region where air quality/emission source relationships are
poorly understood.

A third method of relating air quality to emissions starts
from the characteristics of the aerosol at a receptor site (or
sites). (A receptor site is a measurement point not directly
located in the effluent stream from an emission source.) The
measured properties of the aerosol including total mass, light
extinction or chemical composition are then allocated to the
separate sources contributing to the aerosol by methods such as
those described below. This approach-starting from the
measurement site and working back to the sources-is known as
receptor modeling. The receptor model approach has recently been
reviewed by Gordon (2).

Aside from applications to specific regions or locations, new
developments in receptor modeling have tended to take place in one
of three broad categories: experimental methods, data analysis and

physical theory. The development of novel experimental techniques
has been stimulated by the need for closing mass balances on the
collected aerosol or of measuring concentrations of key trace
substances. Carbon and its compounds have been a particularly
difficult component of the aerosol on which to close mass balances
because of measurement difficulties.

A considerable effort has gone into the application of
statistical techniques to the analysis of aerosol data for the
extraction of source contributions. The use of novel statistical
methods has been stimulated by uncertainties in the data collected
in field measurements and in source characterization; in some
cases not all of the sources are known.

The basic theoretical equation (3) relating source
contributions and chemical composition is a mass balance which
requires no consideration of rate processes. In this paper, the
theory is extended to the resolution of the visibility degrading
components of the aerosol and to chemically reactive families of
chemical compounds. These extensions require new theoretical
analyses which take into account the dynamics of aerosol growth
and chemical kinetics, respectively. The extension to these rate
processes are the subject of this paper.

We start from the chemical element balance (CEB) method (3) of
source resolution as a reference approach although it is not
necessary to all of the discussion which follows.

Chemical Element Balances: Maximum Likelihood Method

The concentration of element i (mass per unit volume of air)
measured at a receptor site is related to the source contributions
by

$$\rho_i = \sum_j c_{ij}\, m_j \qquad\qquad i = 1, 2, \ldots . n \qquad (1)$$

where

c_{ij} = mass fraction of species i in the particulate matter
from source j at the receptor site.

m_j = mass of material from source j per unit volume of air
at the receptor site.

The source concentration matrix c_{ij} should correspond to the

point of measurement. If there is no fractionation between the
source and receptor site, c_{ij} is equal to its value at the
source; this is the assumption usually made.

The goal of the analysis is to determine the source
contributions m_j by inverting Eq. (1). Usually there are more
equations than unknowns, that is, more measured elemental
concentrations ρ_i than source contributions m_j. The values
of m_j for this over-determined system can be estimated using
a least squares method in which the following assumptions are
made:

1. The errors affecting the measurement of ρ_i are
    normally distributed and uncorrelated.

2. The measured values of the source concentrations are
    exact.

3. The set of measurements of ρ_i is the most probable
    set (maximum likelihood principle).

If the first two conditions are met, the probability, P, of
observing a set of measured concentrations between ρ_i ,...ρ_n and
$\rho_1 + d\rho_1$,...$\rho_n + d\rho_n$ is (4):

$$P = \frac{1}{\sigma_{\rho_1} \cdots \sigma_{\rho_n} (\sqrt{2\pi})^n} \exp\left\{ -\frac{1}{2} \sum_{i=1}^{n} \frac{\left(\rho_i - \sum_{j=1}^{P} c_{ij} m_j\right)^2}{\sigma_{\rho_1}^2} \right\} d\rho_1 \cdots d\rho_n \quad (2)$$

where $\sum_{j=1}^{P} c_{ij} m_j$ represents the exact value of ρ_i obtained in
the absence of error in the measurements; σ_{ρ_i} is the standard
deviation in the measurement of ρ_i.

By the third assumption above, the measured set of values of
ρ_i represents the most probable set, which is equivalent to
maximizing P in Eq. (2).

The value of P is maximized by choosing the values of the m_j's which minimize the argument of the exponential function in Eq. (2):

$$X = \sum_{i=1}^{n} \frac{\left(\rho_i - \sum_{j=1}^{P} c_{ij} m_j\right)^2}{\sigma_{\rho_i}^2} \qquad (3)$$

Setting the derivative of X with respect to each m_j equal to zero results in an expression which can be solved for the individual source contributions m_i.

Previous applications of the CEB method have been reviewed by the NAS (5). The method has been applied by Mizohata and Mamuro (6) to separate size ranges of the Osaka (Japan) aerosol collected with a cascade impactor. Watson (7) has extended the CEB method to cases where there are uncertainties of known value in the source concentrations c_{ij} as well as in ρ_i.

Visual Range-Emission Source Relationships
(based on Friedlander, Chap. 11 (1) and Ouimette (8))

The visual range s^* is related to the extinction coefficient by the expression:

$$s^* = \frac{3.912}{b} \qquad (4)$$

The extinction coefficient for an aerosol composed of spherical particles is given by:

$$b = \int_0^{\infty} \frac{\pi d_p^2}{4} K_{ext} (x,m) \, n(d_p) \, d(d_p) \qquad (5)$$

where K_{ext} is the particle extinction cross-section, $x = \pi d_p / \lambda$, m = refractive index and λ is the wavelength of the incident light. It is necessary to clarify the nature of the extinction cross-section. If all the particles in the aerosol are spherical

and composed of the same material, there is no ambiguity, and the
extinction cross-section can be obtained from Mie theory for a
homogeneous sphere of a given refractive index. If the particles
originate from different sources and consequently are of different
chemical composition (and refractive index), a strong coagulation
process would be necessary to achieve uniform composition at the
particle level. The time to achieve this degree of coagulation is
usually too great compared with atmospheric residence times.
Hence such aerosols, internal mixtures, are usually not good
representation for a polluted atmosphere containing particles from
many different sources.

In the general case, individual particles have differing
compositions and refractive indices and to take this into account
in detail is not possible from a practical point of view. To
allow for a variation of refractive index, a convenient model is
that of a mixture of aerosols from the several sources, each with
its own extinction cross-section. The particles are assumed not
to coagulate so that the aerosol is not mixed on the individual
particle basis. Such an aerosol is known as an external mixture.
This model would also be applicable, approximately, to an aerosol
mixture whose particles are growing in size by gas-to-particle
conversion.

Equation (5) can then be applied to each component of the
aerosol. For component i of the mixture, this expression can be
rewritten as follows:

$$b_i = \frac{3}{2} m_i \int_{-\infty}^{+\infty} \frac{K_i}{\rho_i d_p} \frac{dM_i}{\rho_i m_i \, d \log d_p} \, d \log d_p \qquad (6)$$

where $dM_i = \rho_i \frac{\pi d_p^3}{6} n_i(d_p) \, d(d_p)$ is the mass of material from
source i in the particle size range between d_p and $d + d(d_p)$,
ρ_i is the density of the material from source i and m_i is
the total mass of material from source i .

The extinction coefficient for the mixture is then given by

$$b = \sum_i b_i = \sum_i \gamma_i m_i \qquad (7)$$

where γ_i the extinction coefficient per unit mass for each component is given by the expression:

$$\gamma_i = \frac{3}{2} \int_{-\infty}^{+\infty} \frac{K_i}{d_p} \frac{dM_i}{\rho_i m_i \, d \log d_p} \, d \log d_p \tag{8}$$

There are at least two ways to evaluate the coefficients γ_i : They can be calculated from theory if the extinction cross-section and mass distributions are known. Calculations of this type have recently been made by Ouimette (8).

The coefficients γ_i can also be obtained empirically by measuring b , the m_j's and then carrying out a regression analysis to determine the best set of values for γ_i (9).

Conditions for Constant γ_i

Values of γ_i vary for different aerosol components, the large values corresponding to the components with the highest extinction coefficients per unit mass of aerosol material. When γ_i is constant, the extinction coefficient is linearly related through the coefficients γ_i to the mass contributions of the various sources; this considerably simplifies analyses relating visibility degradation to source contributions.

Equation (8) shows that γ_i depends primarily on two factors, the extinction cross-section K_{ext} and the normalized mass distribution of the aerosol. For γ_i to be constant at a given location, the normalized mass distribution must be independent of time. Similarly for γ_i to be independent of location, the normalized mass distribution must not vary from place to place. Recent measurements by Ouimette (8) with the low pressure impactor at a site in China Lake, California have shown that normalized mass distributions for certain aerosol components varied relatively little over a period of months at the site where measurements were made.

It is of interest to examine theoretically the circumstances under which values of γ_i remain constant. A simple case is that of pure dilution of an aerosol; the normalized mass distribution in this case does not change.

When the size distribution changes by gas-to-particle
conversion, the following analysis is applicable: Returning to the
fundamental relationship, Eq. (5), it is clear that another
condition under which γ_i is constant is given by

$$\frac{n_i(d_p,t)}{v_i(t)} = f_i(d_p) \tag{9}$$

where f_i is a function only of particle diameter and $v_i(t)$ is
the total aerosol volume. This is equivalent to making the
normalized mass distribution a function only of particle diameter
and not of time.

Rewriting Eq. (9),

$$n_i(d_p,t) = v_i(t)\, f_i(d_p) \tag{10}$$

we see that $n_i(d_p,t)$ is a separable function of d_p and t. If
the only mechanism for change in $n_i(d_p,t)$ is gas-to-particle
conversion, the dynamic equation for $n_i(d_p,t)$ is given by ($\underline{1}$).

$$\frac{\partial n_i}{\partial t} + \frac{\partial n_i q}{\partial d_p} = 0 \tag{11}$$

where $q = \dfrac{d(d_p)}{dt}$ is the particle growth law whose form depends
on the mechanism of gas-to-particle conversion. The growth law
can usually be represented by a separable function:

$$q = g(t)\, h(d_p) \tag{12}$$

For diffusional growth, $g(t)$ is proportional to the
concentration of the diffusing species and $h \sim d_p^{-1}$ ($\underline{1}$).

Substituting Eqs. (10) and (12) in Eq. (11) and rearranging,
the result is

$$\frac{dv_i}{gv_i dt} + \frac{df_i h}{f_i d(d_p)} = 0 \tag{13}$$

The first term is a function only of t and the second of d_p. This can only be the case if both are equal to constants:

$$\frac{dv_i}{gv_i dt} = k \tag{14a}$$

and

$$\frac{df_i h}{f_i d(d_p)} = -k \tag{14b}$$

Integrating Eq. (14b) the result is

$$f_i = \frac{A}{h} \exp \left[-k \int d(d_p)/h\right] \tag{15}$$

where A is a constant.

For diffusion-limited particle growth, $h = B/d_p$ where B is a constant. Substituting in Eq. (15) then gives

$$f_i = \frac{Ad_p}{B} \exp \left[-k\, d_p^{\,2}/2B\right] \tag{16}$$

This is the special form which f_i in Eq. (10) must take for γ_i to be constant during diffusion-limited growth. It corresponds to a unimodal particle size distribution characterized by the two constants k/B and A/B.

Thus the use of a linear theory <u>with</u> <u>constant</u> <u>coefficients</u> to relate the extinction coefficient to source mass contributions can be justified for certain aerosol growth mechanisms.

<u>Chemical Species Balances: Decay Factors for Reactive Species</u>

The CEB method can be extended to families of chemical compounds, such as polycyclic aromatic hydrocarbons (PAH), while taking chemical reaction into account. Formally this can be done by writing chemical species balances in a somewhat different fashion from the CEB formulation:

$$\rho_i = \sum_j \alpha_{ij}\, x_{ij}\, m_{1j} \tag{17}$$

where

m_{1j} = concentration of reference species 1 originating from source j , measured at the receptor site, $\mu g/m^3$.

x_{ij} = ratio of mass of species i to the reference species 1 in the emissions from source j , dimensionless.

α_{ij} = decay factor = fraction of species i emitted from source j remaining in the aerosol at the receptor site, dimensionless.

It is convenient to select as the reference species, a non-reacting compound for which $\alpha = 1$, if such a species is present. This formulation, Eq. (17) is used because PAH concentrations are usually not reported on the mass fraction basis (mass of PAH per unit mass of particulate matter), but on an absolute basis (mass per unit volume of gas, $\mu g/m^3$).

Once emitted, individual compounds may react chemically in three post-emission stages: First, while suspended in the atmosphere before being sampled, the compound may react in the presence of solar radiation and various reactive species such as hydroxyl radicals and ozone. For example, the average residence time in the Los Angeles atmosphere of a parcel of air is of the order of ten hours.

The second stage occurs during sampling while aerosol material is deposited in the sampling device, usually a filter but sometimes an impactor. In this stage, the deposited material is exposed to those gaseous species which have survived passage through the sampling line, but not to solar radiation. The second stage may last from a day to a few weeks.

The third stage covers the period during which the sample is stored, before chemical analysis. The decay factor calculated from the chemical analysis of a deposited sample is averaged over the three stages.

Relating Atmospheric Decay Factors to Laboratory Rate Studies

The fraction of species i remaining after chemical reaction in the atmosphere was determined in the previous section by comparison of concentrations at the source and in the atmosphere, with the assumption that one species is non-reacting. The chemical reactivity of these species can also be determined from laboratory studies under controlled conditions of radiation, and gas and aerosol phase composition. (Indeed, the non-reactive species are best identified from laboratory data). It is of interest to relate field data for α_{ij} to laboratory reaction rate parameters. For this purpose, it is necessary to develop a suitable chemical reactor model for the atmospheric reaction processes.

We consider only the relatively simple case of a first order chemical decay process. This is consistent with the way laboratory data for PAH reactions are reported (10).

The chemical species balance method can be extended to first-order chemical decay processes as follows:

The rate of decay of species i of the aerosol is given by

$$-\frac{d\rho_i}{dt} = k_i \rho_i \tag{18}$$

where k_i is the first order reaction rate constant for species i . Integrating from the time of release from the source t = 0 to the time of collection at the receptor t = τ:

$$\rho_i = \rho_{io} \, e^{-k_i \tau} \tag{19}$$

where the subscript o refers to the concentration at the source.

In general for any sample taken at a receptor site, there will be a residence time distribution g (τ) :

$$df = g(\tau)d\tau \tag{20}$$

where df is the fraction of the material sampled with atmospheric residence times between τ and τ + dτ .

The amount of species i remaining at the sampling point is obtained by averaging over all residence times:

$$\bar{\rho}_i = \rho_{io} \int_0^\infty g(\tau) \, e^{-k_i \tau} \, d\tau \tag{21}$$

Further analysis of the problem requires the introduction of the residence time distribution g(τ) . As a first approximation, the atmospheric region of interest can be assumed to behave like a

continuous stirred tank reactor (CSTR). Reactants introduced into
a CSTR immediately reach a uniform concentration determined by
reactor volume and feed flow rate. The concentrations of the
species in the streams leaving the reactor are equal to the
reactor concentrations. This is equivalent to a simple
atmospheric box model with a fixed inversion lid (11).

The residence time distribution for a CSTR is given by (12):

$$g(\tau) = \frac{1}{\theta} e^{-\tau/\theta} \tag{22}$$

Here θ , the average residence time, is given by

$$\theta = \frac{V}{q} \tag{23}$$

where V is the reactor volume and q is the flow rate.

Substituting in Eq. (21) and integrating, the result is

$$\bar{\rho}_i = \frac{\rho_{io}}{1 + k_i \theta} \tag{24}$$

and the decay factor is given by

$$\alpha_i = \frac{\bar{\rho}_i}{\rho_{io}} = (1 + k_i \theta)^{-1} \tag{25}$$

The decay factor is equal to unity for a non-reacting component
($k_i = 0$) and approaches zero fcr a component which reacts rapidly
($k_i \to \infty$).

Decay Factors for Ambient PAH

The approach discussed above can be applied to the ambient
polycyclic aromatic hydrocarbons (PAH). Studies designed to take
into account all of the factors discussed in the previous sections
have not been conducted, but data available in the literature can
be used to illustrate the method.

Detailed measurements of atmospheric concentrations of
fourteen PAH were made by Gordon and Bryan (13) and Gordon (14) at
various locations in Los Angeles. They assumed that the

concentrations at a freeway site were dominated by automobile emissions. Assuming also that coronene was emitted only by automobiles, they estimated the non-auto PAH in the total measured at a site near a concentration of petroleum refineries and chemical plants, with only moderate traffic.

More recently, Duval (15) estimated decay factors for the ambient PAH in Los Angeles using the same data. Following Gordon and co-workers, he assumed that PAH concentrations at a site in downtown Los Angeles were automobile dominated; a site near to and downwind from the refineries had both automobile and refinery contributions. Using lead as a tracer for the automobile component at the refinery site, Duval determined the PAH emission profile for the refinery emissions.

To determine the decay factors, α_i, Duval used the data of Grimmer and Hildebrant (16) for automobile emissions of PAH. These investigators measured the emissions of fourteen PAH by twenty of the most common car models in Germany based on their frequency of registration. Some of the PAH were the same as those measured by Gordon (14). The car models were not identified by make. Five cars of each model type, a total of 100, were tested using a standard test which simulated city driving. The same gasoline fuel was used in each case.

Although the absolute amounts of the individual PAH varied from car to car and model to model, the relative amounts were independent of car model, within experimental error. To calculate decay factors, Duval assumed that the Los Angeles vehicle PAH emission profile is the same as that for the German fleet. With this strong assumption, it was possible to calculate the decay factors shown in Table 1. It is not known which of the three post-emission reaction stages, discussed in a previous section, was most important in determining the decay factors. There was, however, no clear seasonal variation in decay factors and they were averaged over all of the data. The PAH emission concentrations used in the calculation of the decay factors (16) are shown in Table 2 along with the calculated emissions from the refineries based on the decay factors of Table 1.

If it is assumed that the first (atmospheric) stage controls,

Table 1.   PAH Decay Factors [a]
(reference 15)

PAH	Decay Factor
BaP	0.48 ± 0.21
BeP	1.04 ± 0.29[b]
BFL	0.98 ± 0.26
BghiP	0.83 ± 0.19
ANTH	0.21 ± 0.11

[a] Arithmetic mean within 68% confidence.
[b] BeP was assumed a stable species in the source resolution analysis.

Nomenclature (Tables 1, 2, and 3):

PAH   :   polycyclic aromatic hydrocarbons
BFL   :   benzofluoranthenes
BaP   :   benzo(a)pyrene
BeP   :   benzo(e)pyrene
INP   :   Indeno(1,2,3-cd)pyrene
COR   :   coronene
BghiP: benzo(ghi)perylene
ANTH :   Anthanthrene

Table 2. Source Concentrations for PAH

Source of Emmissions	BFL	BaP	PAH/Bep BghiP	ANTH	COR
Automobiles (Grimmer et al., 1975)	1.08 ± 0.36	0.98 ± 0.36	3.10 ± 0.87	0.57 ± 0.27	1.96 ± 0.46
Refineries [a]	1.43	3.85	2.46	2.12	0.23[b]

[a] All PAH were corrected for their decay using the decay factors shown in Table 1.

[b] Based on small coronene concentrations, and the error is expected to be large.

Table 3. Comparison of rates of degradation of selected PAH[a]

Reference	System	Adsorbent	Reactant	Light intensity	Rate constants hr^{-1}	
Katz et al[b]	Semi-static system	Cellulose	hν and air	quartzline lamp, 46 cm above the samples	BaP BeP	0.13 0.03
			hν, air and 0.2 ppm O_3	$(9.4\ 10^{13}$ quanta/s/ $cm^2)$	BaP BeP	1.19 0.13
This work	CSTR	Atmospheric samples taken in Los Angeles (Site 1). The residence time estimated from lead emissions and lead atmospheric concentrations was 11 hours.			ANTH BeP BeP BFL BghiP	0.372 0.098 0 0.002 0.019
					ANTH BaP BeP BFL BghiP	0.142 0.067 0 0.002 0.017

[a] reference 15.

[b] Rate constants shown were obtained from the corresponding PAH half-lives reported by the authors (10).

first order reaction rate constants can be calculated from
Eq. (25) for CSTR theory. Values calculated in this way by Duval
are shown in Table 3. These results were compared with data
reported in the literature for PAH degradation in laboratory
studies with simulated atmospheres. Best agreement was found with
the data of Katz et al (10) for exposure to simulated sunlight in
air in the absence of O_3.

Summary and Conclusions

Extensions of receptor modeling theory to light extinction and
to reactive chemical compounds require introduction of rate
process concepts.

Light extinction coefficients per unit mass of chemical
constituent are constant for growing aerosols for certain forms of
the growth laws and particle size distributions. Constant
coefficients simplify source allocation analyses for visibility
degradation.

The CEB method can be extended to chemically reactive species
by introducing decay factors into the mass balances for the
chemical species. The decay factors can be evaluated from data
for the composition of emissions and of the ambient aerosol. They
can be related to first order reaction rate coefficients measured
in the laboratory by means of an appropriate atmospheric model.

Finally, receptor modeling offers a useful theme around which
to organize aerosol characterization studies. Large scale field
studies are expensive, and they also tend to be diffuse. Data
requirements for source resolution can be used to select the
chemical and physical properties of the aerosol to be measured
both at receptor sites and at sources.

Acknowledgements
This work was supported in part by EPA Grant No. R806404-02-1.
The contents do not necessarily reflect the views and policies of
the Environmental Protection Agency.

Literature Cited

1. Friedlander, S.K., (1977) SMOKE, DUST AND HAZE:
   FUNDAMENTALS of AEROSOL BEHAVIOR, Wiley-Interscience, New
   York.

2. Gordon, G.E., (1980) "Receptor Models," Environ.
   Sci. Technol., 14, 792-800.

3. Friedlander, S.K., (1973) "Chemical Element Balances and
   Identification of Air Pollution Sources,"
   Environ. Sci. Technol., 7, 235-40.

4. Young, H.D., (1962) STATISTICAL TREATMENT of EXPERIMENTAL
   DATA, McGraw Hill, New York, Ch. 4.

5. NAS, (1980) CONTROLLING AIRBORNE PARTICLES, National
   Academy of Sciences, Washington, D.C., Ch. 3.

6. Mizohata, A.and Mamuro, T., (1979) "Chemical Element
   Balances in Aerosol Over Sakai, Osaka", Annual Report of
   the Radiation Center of Osaka Prefecture, 20, 55-69.

7. Watson, J.G., Jr., (1979) "Chemical Element Balance
   Receptor Model Methodology for Assessing the Sources of
   Fine and Total Suspended Particulate Matter in Portland,
   Oregon," Thesis, Ph.D., Department of Chemistry, Oregon
   Graduate Center, Beaverton, OR.

8. Ouimette, J.R., (1980) "Chemical Species Contributions to
   the Extinction Coefficients," Thesis, Ph.D.,
   Environmental Engineering, California Institute of
   Technology, Pasadena, CA.

9. White, W.H., Robert, P.T., (1980) "On the Nature and
   Origins of Visibility-Reducing Aerosols in the Los
   Angeles Air Basin," in THE CHARACTER and ORIGINS of SMOG
   AEROSOLS, G.M. Hidy et al., (Editors) Wiley-Interscience,
   New York, Part IV, 715-753.

10. Katz, M., Chan, C., Tosine, H., Sakuma, T., (1979)
    "Relative Rates of Photochemical and Biological Oxidation
    (in vitro) of Polynuclear Aromatic Hydrocarbons," in
    POLYNUCLEAR AROMATIC HYDROCARBONS, P.W. Jones and
    P. Leber (Editors), Ann Arbor Science Publishers, Inc.,
    Ann Arbor, MI, 171-89.

11. Tennekes, H., (1976) "Observations on the Dynamics and
    Statistics of Simple Box Models with a Variable Inversion
    Lid," in Third Symposium on Atmospheric Turbulence,
    Diffusion and Air Quality, American Meteorological
    Society, Boston, MA, 397-402.

12. Denbigh, K.G.and Turner, J.C.R., (1971) CHEMICAL REACTOR
    THEORY: AN INTRODUCTION, 2nd Ed., Cambridge University
    Press, Cambridge, ENGLAND.

13. Gordon, R.J., Bryan, R.J., (1973) "Patterns in Airborne
    Polynuclear Hydrocarbons Concentrations at Four Los
    Angeles Sites", Environ. Sci. Technol., 7, 1050-3.

14. Gordon, R.J., (1976) "Distribution of Airborne Polycyclic
    Aromatic Hydrocarbons Throughout Los Angeles",
    Environ. Sci. Technol., 10, 370-3.

15. Duval, M.M., (1980) "Source Resolution Studies of Ambient
    Polycyclic Aromatic Hydrocarbons in the Los Angeles
    Atmosphere: Application of a Chemical Species Balance
    Method with First Order Chemical Decay", Thesis, Master
    of Science in Engineering, UCLA.

16. Grimmer, G., Hildebrandt, A., (1975) "Investigations on
    the Carcinogenic Burden by Air Pollution in Man XII.
    Assessment of the Contribution of Passenger Cars to Air
    Pollution by Carcinogenic Polycyclic Hydrocarbons",
    Zbl.Bakt.Hyg., I.Abt.Orig.B 161, 104-24.

RECEIVED April 24, 1981.

The Application of Factor Analysis to Urban Aerosol Source Resolution

PHILIP K. HOPKE

Institute for Environmental Studies, University of Illinois, Urbana, IL 61801

Among the multivariate statistical techniques that have been used as source-receptor models, factor analysis is the most widely employed. The basic objective of factor analysis is to allow the variation within a set of data to determine the number of independent causalities, i.e. sources of particles. It also permits the combination of the measured variables into new axes for the system that can be related to specific particle sources. The principles of factor analysis are reviewed and the principal components method is illustrated by the reanalysis of aerosol composition results from Charleston, West Virginia. An alternative approach to factor analysis, Target Transformation Factor Analysis, is introduced and its application to a subset of particle composition data from the Regional Air Pollution Study (RAPS) of St. Louis, Missouri is presented.

There has recently been a surge of interest in the development and application of techniques that permit the identification and quantitative apportionment of sources of urban aerosol mass. Among these techniques are various forms of a statistical method called factor analysis. Several forms of factor analysis have been applied to the problem of aerosol source resolution. These different forms provide several different frameworks in which to examine aerosol composition data and interpret it in terms of source contributions.

In an aerosol sampling and analysis program, a large number of samples, n, are taken and analyzed for many elemental concentrations. Thus, a large matrix of data is obtained. We can think of plotting the values obtained in a multidimensional space.

0097-6156/81/0167-0021$07.50/0
© 1981 American Chemical Society

If we have determined m elements in n samples, we could display
these results as n points in an m-dimensional space. However,
because some of the elements are emitted by the same source of
particles, their concentrations will be related.
    To illustrate this idea, suppose that we have two sources of
particles, an iron foundry and automobile emissions, and that we
measure three elemental concentrations, iron, lead, and bromine.
If we plot concentrations in a three dimensional space, a point
represents a sample as shown in Figure 1. In fact, the space
necessary to show all of the points is really only two
dimensional, since the amount of lead and bromine are directly
interrelated. By a simple axis rotation shown in Figure 2, it can
be seen that all of the points lie in a plane defined by the iron
axis and a line that defines the lead-bromine relationship. For
more complex systems, factor analysis will help to identify the
true dimensionality of the system being studied and permit the
determination of these interelemental relationships. With this
information, it is then possible to determine the mass
contribution of the particle sources to the total observed mass.

Statistical Background

    In determining the quantity of a particular element or
compound in a specific sample at a definite time, the investigator
has randomly removed a sample from a distribution of materials
present in the environment. Then, by taking enough samples, the
distribution of that particular variable in that kind of sample
can be described by several parameters commonly used for that
purpose including the mean value of the jth variable

$$x_i = 1/n \sum_{j=1}^{n} (x_{ij}) \qquad (1)$$

and the second moment of the distribution or the variance

$$s_i^2 = \left[1/(n-1)\right] \sum_{j=1}^{n} (x_{ij} - x_j)^2 \qquad (2)$$

where n is the number of samples examined. The standard deviation
is simply the square root of the sample variance.
    For some statistical procedures, it is necessary to remove the
effects of using different metrics in describing the various
variables, so the variables are put in standard form. First, the
deviation is calculated by subtracting the mean value from each
sample value.

$$d_{ij} = x_{ij} - x_j \qquad (3)$$

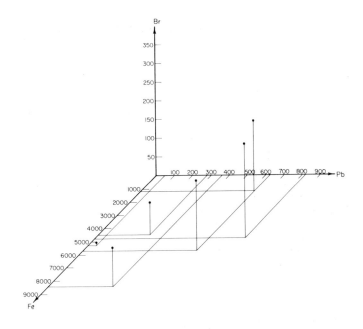

Figure 1.   *Artificial aerosol composition data plotted for automobile and steel plant emissions*

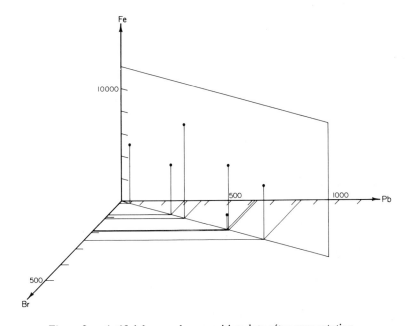

Figure 2.   *Artificial aerosol composition data after axes rotation*

The standardized variable, z_{ij}, can be calculated by dividing the deviation by the standard deviation

$$z_{ij} = \frac{d_{ij}}{s_j} = \frac{x_{ij} - x_j}{s_j} \tag{4}$$

The standardized value then has a mean value of zero and a standard deviation of unity, and thus, all standardized variables have a mean value of zero and a standard deviation of 1.

The initial step in the analysis of the data generally requires the calculation of a function that can indicate the degrees of interrelationship that exist within the data. Functions exist that can provide this measure between either the variables when calculated over all of the samples or between the samples calculated over the variables. The most well-known of these functions is the product-moment correlation coefficient. To be more precise, this function should be referred to as the correlation about the mean. The "correlation coefficient" between two variables, x_j and x_k over all n samples is given by

$$c_{ik} = \frac{\Sigma (x_{ij}-x_i)(x_{kj}-x_k)}{(\Sigma (x_{ij}-x_i)^2 \ \Sigma (x_{kj}-x_k)^2)^{1/2}} \tag{5}$$

utilizing the standardized variables eq. (5) simplifies to

$$c_{ik} = (1/n) \sum_{j=1}^{n} z_{ij} z_{kj} \tag{6}$$

There are several other measures that can also be utilized. These measures include the covariance about the mean

$$co_{ik} = (1/n) \sum_{j=1}^{n} d_{ij}d_{kj} \tag{7}$$

covariance about the origin

$$co'_{ik} = (1/n) \sum_{j=1}^{n} x_{ij}x_{kj} \tag{8}$$

and the correlation about the origin

$$c'_{ik} = \frac{\Sigma \ x_{ij} \ x_{kj}}{(\Sigma x_{ij}^2 \ \Sigma x_{kj}^2)} \tag{9}$$

The matrix of all of either the correlations or covariances or the dispersion matrix can be obtained from the original or transformed data matrices. The data matrices contain the data for the m variables measured over the n samples. The correlation about the mean is given by

$$(C_m) = (Z)(Z)^t \tag{10}$$

where $(Z)^t$ is the transpose of the standardized data matrix (Z). The correlation about the origin

$$(C_o) = (Z^o)(Z^o)^T = (XV)(XV)^t \tag{11}$$

where $z^o_{ij} = x_{ij}/(\sum\limits_{j=1}^{n} x_{ij}^2)^{1/2}$

which is a normalized variable still referenced to the original variable origin and (V) is a diagonal matrix of the square root of the sums of the squares of the variables. The covariance about the mean is given as

$$(Co_m) = (D)(D)^t \tag{12}$$

where $(D)^t$ is the transpose of the matrix of deviations from the mean calculated using equation 3. The covariance about the origin is

$$(Co_o) = (X)(X)^t \tag{13}$$

the simple product of the data matrix by its transpose. As written these product matrices would be of dimension m by m and would represent the pairwise interrelationships between variables. If the order of the multiplication is reversed, then the resulting n by n dispersion matrices contain the interrelationships between samples.

There has been some discussion in the literature ([1-3]) regarding the relative merits of these functions to reflect the total information content of the data. Rozett and Peterson ([1]) argue that since many types of physical and chemical variables have a real zero, the correlation and covariance about the mean remove the information regarding the location of the original origin by only including differences from the variable mean. The normalization that is made in calculating the correlation from the covariances caused each variable to have an identical weight in the subsequent analysis. In mass spectrometry where the variables consist of the various m/e values observed for the fragments of a molecule, the normalization represents a loss of information because the metric is the same for all of the m/e values. In environmental studies where concentrations range from trace species at the sub part-per-million level to major constituents at the percent level, the use of covariance may weight the major constituents too heavily in the subsequent analyses. The choice of function depends heavily on the underlying assumptions made concerning the nature of the parameters being measured.

Besides having a convenient measure of the interrelationship between two variables, it is also useful to develop procedures to describe the relationships betweeen samples so that subsequently the samples can be grouped according to how similar or dissimilar they are to one another. One set of possible functions to describe the relationship between samples is the correlation and covariance function defined in the equations 6 to 9 with the meaning of the indexes changed so that j and k refer to different samples and the summations are taken over the n variables in the system. The use of such functions will be explored later in this chapter.

Principal Components And Common Factor Analysis

Basic Concepts. The goal of factor and components analysis is to simplify the quantitative description of a system by determining the minimum number of new variables necessary to reproduce various attributes of the data. Principal components analysis attempts to maximally reproduce the variance in the system while factor analysis tries to maximally reproduce the matrix of correlations. These procedures reduce the original data matrix from one having m variables necessary to describe the n samples to a matrix with p components or factors (p<m) for each of the n samples. In addition to permitting a parsimonious representation of data, it also assists in the identification of the nature of each of the factors.

In both component and factor analysis, the properties of the system being observed are assumed to be linearly additive functions of the contribution from each of the m causalities that actually govern the system. For example, for airborne particles, the amount of particulate lead in the air could be considered to be a sum of contributions from several sources including automobiles, incinerators and coal-fired power plants, etc.

$$M_T(Pb) = M_1(Pb) + M_2(Pb) + M_3(Pb) + \ldots \tag{14}$$

Where M_T is the observed total particulate lead per unit volume, $M_1(Pb)$ is the amount per unit volume from source 1, $M_2(Pb)$ is the amount per unit volume from source 2, and so forth for all of the m possible sources.

The individual source contributions can generally be thought of as a product of two cofactors; one that gives the amount of lead in particles emitted by that particular source and the other which given the amount of particulate matter per unit volume present that can be attributed to that source. Thus, the $M_i(Pb)$'s can be rewritten as

$$M_i(Pb) = a_i(Pb) F_i \tag{15}$$

Expanding this approach to all of the m variables that have been measured for the n samples gives

$$x_{ij} = a_{i1}F_{1j} + a_{i2}F_{2j} + \ldots a_{ip}F_{2p} \tag{16}$$

where x_{ij} is the value of the ith variable for the jth sample. In the usual factor analysis approach, however, the equation is usually written for the standardized variable as defined in equation 4.

$$z_{ij} = a'_{i1}F'_{1j} + a'_{i2}F'_{2j} + \cdots + a'_{ip}F'_{pj} \qquad (17)$$

The advantage of utilizing the standardized form of the variable is that quantities of different types can be included in the analysis including elemental concentrations, wind speed and direction, or particle size information. With the standardized variables, the analysis is examining the linear additivity of the variance rather than the additivity of the variable itself. The disadvantage is that the resolution is of the deviation from the mean value rather than the resolution of the variables themselves. There is, however, a method to be described later for performing the analysis so that equation 16 applies. Then, only variables that are linearly additive properties of the system can be included and other variables such as those noted above must be excluded. Equation 17 is the model for principal components analysis. The major difference between factor analysis and components analysis is the requirement that common factors have the significant values of a for more than one variable and an extra factor unique to the particular variable is added. The factor model can be rewritten as

$$z_{ij} = a'_{i1}F'_{1j} + a'_{i2}F'_{2j} + \cdots + a'_{ip}F'_{pj} + d_i U_{ij} \qquad (18)$$

where U_{ij} is a factor unique to the ith variable calculated for the jth sample. The unique factor includes both causal factors related only to that variable as well as the error associated with that variable.

In the components analysis, components can be found with a strong relationship to only a single variable. This single variable component could also be considered to be the unique factor. The two models really aim at the same results and the differences are primarily of definition and sematics. The model as expressed in equation 17 can be expressed in matrix notation as

$$(Z) = (A')(F') \qquad (19)$$

where (Z) is the m x n data matrix, (A') is the m x p factor loading matrix and (F') is the p x n factor score matrix. Factor analysis must permit determination of the number of terms required, p, and of one of the two matrices on the right hand side of equation 19.

Both component and factor analysis as defined by equations 17 and 18 aim at the identification of the causes of variation in the system. The analyses are performed somewhat differently. For the principal components analysis, the matrix of correlations defined by equation 10 is used. For the factor analysis, the diagonal elements of the correlation matrix that normally would have a value of one are replaced by estimates of the amount of variance that is within the common factor space. This problem of separation of variance and estimation of the matrix elements is discussed by Hopke et al. (4).

The next step is the diagonalization of the matrix. This process
compresses the information content of the data set into a limited number
of eigenvalues while the remaining eigenvectors would contain the error
content of the data. In theory, there should be a sharp distinction
between non-zero and zero values. In practice, the choice of the number
of eigenvectors to be retained is often extremely difficult since there
are generally a number of small but non-zero values. Often it is
helpful to plot the eigenvalues as a function of eigenvalue number and
look for sharp breaks in the slope of the line that often can provide an
indication of the point of separation (5). A number of commonly used
methods for determining the number of retained factors have been
reviewed (3). In general the average error appears to be the most
useful criterion where the number of factors is determined by the number
that are required to reproduce the original data within the average
root-mean-square uncertainty of the data.

After the number of factors have been determined, it is necessary to
interpret the factors as physically real sources. For the applications
of this approach to aerosol source identification (4,6-10), the reduced
size matrix of eigenvectors was rotated in such a way as to maximize the
number of values that are zero or unity. This rotation criterion,
called "simple structure" is described in the appendix of reference 4.
A Varimax rotation (11) is often used to achieve it. However, simple
structure may not be the most useful criterion for environmental source
resolution since an element may be present in an aerosol sample because
of its emission by several sources. The variance should, therefore, be
spread over several factors rather than concentrated in one.

Prior Applications. The first application of this traditional factor
analysis method was an attempt by Blifford and Meeker (6) to interpret
the elemental composition data obtained by the National Air Sampling
Network(NASN) during 1957-61 in 30 U.S. cities. They employed a
principal components analysis and Varimax rotation as well as a
non-orthogonal rotation. In both cases, they were not able to extract
much interpretable information from the data. Since there is a very
wide variety of sources of particles in 30 cities and only 13 elements
measured, it is not surprising that they were unable to provide much
specificity to their factors. One interesting factor that they did
identify was a copper factor. They were unable to provide a convincing
interpretation. It is likely that this factor represents the copper
contamination from the brushes of the high volume air samples that was
subsequently found to be a common problem (12).

Hopke, et al. (4) and Gaarenstroom, Perone, and Moyers (7) used the
common factor analysis approach in their analyses of the Boston and
Tucson area aerosol composition, respectively. In the Boston data, for
90 samples at a variety of sites, six common factors were identified
that were interpreted as soil, sea salt, oil-fired power plants, motor
vehicles, refuse incineration and an unknown manganese-selenium source.
The six factors accounted for about 78% of the system variance. There
was also a high unique factor for bromine that was interpreted to be
fresh automobile exhaust. Large unique factors for antimony and
selenium were found. These factors may possibly represent emission of
volatile species whose concentrations do not covary with other elements
emitted by the same source.

In the study of Tucson (7), at each site whole filter data were analyzed separately. They find factors that are identified as soil, automotive, several secondary aerosols such as $(NH_4)_2SO_4$ and several unknown factors. They also discovered a factor that represented the variation of elemental composition in their aliquots of their neutron activation standard containing Na, Ca, K, Fe, Zn, and Mg. This finding illustrates one of the important uses of factor analysis ; screening the data for noisy variables or analytical artifacts.

Gatz (8) applied a principal components analysis to aerosol composition data for St. Louis, Mo taken as part of project METROMEX (13-14). Nearly 400 filters collected at 12 sites were analyzed for up to 20 elements by ion-exchange x-ray fluorescence. Gatz used additional parameters in his analysis including day of the week, mean wind speed, percent of time with the wind from NE, SE, SW, or NW quadrants or variable, ventillation rate, rain amount and duration. At several sites the inclusion of wind data permitted the extraction of additional factors that allowed identification of specific point sources.

Sievering and coworkers (9) have made extensive use of factor analysis in their interpretation of midlake aerosol composition and deposition data for Lake Michigan.

Specific Example. Lewis and Macias (10) have used a principal components analysis on size fractionated aerosol composition data from Charleston, West Virginia. They made the analysis on both coarse and fine samples combined into a single data set and resolved four factors: soil (with some automotive contamination), ammonium sulfate, automotive emissions, and a mixed anthropogenic source. They were unable to separate a coal-combustion source despite its apparent importance as indicated by a high average arsenic concentration of 26 ng/m^3 in the fine fraction. However, they excluded arsenic from the factor analysis because of the inconsistencies in arsenic values obtained from five simultaneously operating samplers. Because of the ability of factor analysis to sort out the sources of variance, it would be useful to observe if the sampling and analysis variance could be separated from the variance resulting from source variation.

It may be possible to obtain additional information from the data of Lewis and Macias by extending the analysis that they performed. A more complete resolution of the sources might be possible if the fine- and coarse- sized particle fractions are separately analyzed. A reanalysis can be made starting from the correlation matrix they report. The eigenvalues for the separated fine and coarse fractions are given in table 1. Lewis and Macias (10) have used an arbitrary cutoff value of unity to decide how many factors to retain. Convincing arguments have been made against the use of this criterion (15) and it is recommended that it not be adopted as the only criterion employed.

An alternative approach for selecting the number of retained factors may be found by examing the partition of variance after the orthogonal rotation. It can be argued that a factor with a variance of less than one contains less information than did one of the original variables. However, since the objective of the rotation is to redistribute the variance from the artificially compressed state that results from the matrix diagonalization, it appears to be useful to examine a number of solutions with differing numbers of retained factors. The rotated solutions that contain factors with total variance less than one can then be rejected. For this example, the fine fraction results yield eigenvalues of 7.35, 2.04, 1.28, 0.81, 0.49, and 0.39 for the first six factors. On the bais of the eigenvalue of one criterion, only three factors would be retained. However, if the 4,5, and 6 factor solutions are examined after a Varimax rotation, the variances for the six factors are 3.27, 3.06, 2.23, 2.11, 1.24, and 0.45. The sixth factor has no loading above 0.42 and this factor appears to contain no useful information. For five factors, all of the factors have a variance greater than one and, therefore, this solution was chosen. Five factors is also the choice for the coarse fraction data as well. The Varimax rotated results are presented in Tables 2 and 3.

For the fine fraction, the first factor clearly represents $(NH_4)_2SO_4$ as Lewis and Macias showed was a major contributor to the fine particle mass. The next factor is strongly associated with iron, calcium, and potassium and is attributed to fine particle soil even though it does have only a moderate silicon loading. The third factor has a higher silicon value and higher values for zinc, selenium and strontium. If it were simply a high zinc factor, it might have been attributable to the zinc contamination from the pump motor described by Lewis and Macias. Since it also includes silicon, selenium and strontium, it may represent coal-fired power plant ash. The fourth factor is clearly automotive emissions. The low value for lead may arise from the interference in the lead determination by the presence of arsenic. The final factor has moderate loadings for carbon and strontium and somewhat lower values for silicon and lead. It is not clear what type of source this factor represents although it may indicate an interference between the Sr L x ray with the Si K x ray intensity that has not been properly calculated.

For the coarse fraction, the first factor contains the majority of the variance and represents soil. There are high loadings for Al, Si, K, Ca, Ti, Fe and Sr. The second factor has high values for bromine and lead. This factor can be identified as motor vehicle exhaust. It does not have a large loading for mass. It would be expected that most of the motor vehicle mass would be found in the fine particle fraction. The third factor has a high loading for carbon, the fourth a high value for nitrogen, and the fifth has a high value for sulfur. These factors indicate that these elements do not covary with other elements. There are no correlation coefficients between these elements and any of the others that are greater than 0.33. The carbon factor shows a relationship with the total coarse particle mass. The source of the carbon factor is not clear. Carbonate levels would be expected to be small and covary with calcium. Humic materials in soils should vary with the elements found in the first factor. This factor may account for pollen since these samples were taken in late summer. Without a wider profile of elements, it is difficult to be more specific. The nitrogen and sulfur factors may represent the higher uncertainties in these determinations than for the fine particle fraction where they were in greater abundance. Lewis and Macias indicate that there were difficulties in the analysis of NH_4^+, NO_3^-, and SO_4^{2-} because of the small amounts present. The point made here is that care must be taken in the application of factor analysis.

This form of factor analysis has the advantages of being able to combine different types of variables in the analysis, of identifying variance in the data that arises from sampling and/or analytical procedure errors, and to provide a prospective of the data without any <u>a</u>

Table 1. Eigenvalues for Factor Analysis of Correlation Matrices of Lewis and Macias(10)

	Fine Fraction			Coarse Fraction	
Number	Eigenvalue	Percent of Total Variance	Number	Eigenvalues	Percent of Total Variance
1	7.3539	56.57	1	8.0782	62.14
2	2.0363	15.66	2	1.6549	12.73
3	1.2792	9.84	3	1.4636	11.26
4	.8093	6.23	4	.7539	5.80
5	.4911	3.78	5	.4863	3.74
6	.3865	2.97	6	.1979	1.52
7	.2176	1.67	7	.1443	1.11
8	.1925	1.48	8	.0750	.58
9	.1031	.79	9	.0612	.47
10	.0847	.65	10	.0372	.29
11	.0283	.22	11	.0303	.23
12	.0167	.13	12	.0142	.11
13	.0009	.01	13	.0029	.02

Table 2. Orthogonally Rotated Factor Matrix
For Fine Fraction Aerosol

Element	1	2	3	4	5	Communality
Mass	.80401	.53753	.04317	.00595	.11566	.9506
C	.32538	.38540	.27853	.51289	.55826	.9067
N	.93182	.09460	.19196	.18688	.08505	.9562
Si	.14505	.53188	.66091	.06791	.42558	.9265
S	.94296	.20758	.03713	.12538	.06226	.9532
K	.12131	.75450	.44304	.29126	.06142	.8689
Ca	.41555	.75257	.27398	.28949	.03917	.8995
Fe	.25148	.93193	.14420	.02478	.12719	.9693
Zn	.01888	.21894	.87835	.24047	-.03313	.8787
Se	.33974	.39535	.70750	.15974	.19940	.8376
Br	.05867	.10013	.31768	.91986	.03476	.9617
Sr	.04830	.01273	.71226	.25545	.61458	.9528
Pb	.40661	.24767	.07840	.69987	.43061	.9081
Variance	3.0683	3.0523	2.7330	1.9721	1.1439	11.9697

Table 3. Orthogonally Rotated Factor Matrix
For Coarse Fraction Aerosol

Element	1	2	3	4	5	Communality
Mass	.59072	.24112	.65619	.21137	-.21149	.9271
C	.09152	.07854	.95699	.09762	.18334	.9271
N	.01945	-.00751	.14641	.97876	-.11491	.9930
Al	.93172	.23120	.18167	.04462	-.05683	.9598
Si	.94557	.25601	.12830	-.03498	-.05162	.9800
S	-.13216	.19330	.10912	-.12604	.95375	.9923
K	.94124	.27771	.09865	-.01809	-.00930	.9732
Ca	.80687	.46949	.18720	-.13533	-.09049	.9330
Ti	.92078	.25257	.10463	-.04673	-.12433	.9402
Fe	.92570	.32771	-.01863	-.05571	-.08429	.9749
Br	.39426	.85928	.10428	.02940	.22489	.9561
Sr	.89285	.10506	.15669	.21470	-.04084	.8805
Pb	.45145	.85422	.12101	-.01638	.07036	.9534
Variance	6.5355	2.1685	1.5354	1.1024	1.0951	12.4370

Table 4. Results of Dimensionality Tests

RAPS Station 112, July and August 1976

Fine Fraction, July 4th and 5th Excluded

Factor	Eigenvalue	Chi Square	Exner	Average % Error
1	87.	7.5	.304	197
2	4.9	2.6	.304	197
3	2.0	0.4	.070	123
4	0.2	0.2	.050	98
5	0.1	0.1	.037	73
6	0.04	0.07	.029	69
7	0.02	0.05	.023	69
8	0.02	0.03	.019	67
9	0.01	0.02	.015	53

Coarse Fraction

Factor	Eigenvalue	Chi Square	Exner	Average % Error
1	96.	3.6	.216	73
2	2.4	1.2	.125	50
3	0.6	0.6	.089	45
4	0.29	0.3	.064	41
5	0.19	0.1	.040	35
6	0.07	0.06	.028	33
7	0.03	0.03	.019	28
8	0.02	0.01	.013	23
9	0.01	0.01	.009	22

priori knowledge of the system under study. This approach has the disadvantages of being unable to quantitatively apportion the aerosol mass among the various sources or to provide the elemental concentration profiles of the sources. In order to overcome these difficulties, an alternative approach to factor analysis has been employed and will subsequently be described.

Target Transformation Factor Analysis

Method Description. By employing a different approach it becomes possible to provide a quantitative apportionment of aerosol mass. The procedure differs from that used above in several important ways. First, the correlation about the origin is employed as the measure of interrelationship. Because the mean value is no longer subtracted from the raw data value, it is possible to produce a result in the form of equation 16.

The second difference is that the correlations between samples are calculated rather than the correlations between elements. In the terminology of Rozett and Peterson (1), the correlation between elements would be an R analysis while the correlation between samples would be a Q analysis. Thus, the applications of factor analysis discussed above are R analyses. Imbrie and Van Andel (16) and Miesch (17) have found Q-mode analysis more useful for interpreting geological data. Rozett and Peterson (1) compared the two methods for mass spectrometric data and concluded that the Q-mode analysis provided more significant informtion. Thus, a Q-mode analysis on the correlation about the origin matrix for correlations between samples has been made (18,19) for aerosol composition data from Boston and St. Louis.

The matrix is diagonalized in the same manner as described above. In the R-mode analysis, the A matrix is obtained and the F matrix is calculated from the data and the A matrix. In the Q-mode analysis the F matrix is initially obtained and the A matrix is calculated.

After diagonalization, the number of factors to be retained is determined. The same problems of determining the number of factors to retain are found in this model. An important area of active research is the exploration of more objective methods of determining the number of factors to be used. In order to illustrate this procedure, an example of the analysis of a subset of data from the Regional Air Pollution Study (RAPS) is used. The set to be analyzed are the data from site 112 (Francis Field on the Washington University Campus) for the months of July and August, excluding July 4 and 5. These samples are excluded because of high contamination of several samples by the Bicentennial fireworks display that could be clearly distinguished in the data set. A factor has been isolated for this source even though it only impacts on three of the one hundred samples included in the analysis. A more detailed description of the data is given by Alpert and Hopke (19).

The tests to determine the number of factors to retain are given in table 4 for both the fine and coarse fractions. For the fine fraction there appear to be three strong sources and two weaker ones. The coarse fraction results do not give a clear indication of the number of factors and parallel analyses with 4 and 5 retained factors were performed until it was found that 4 sources gave the best results.

The major advantage of this form of analysis is that the data have retained their true origin and the columns of the A matrix can be associated with elemental profiles of specific source types. The vectors obtained as a result of the eigenvector analysis are not directly interpretable. At this point it is necessary to rotate the factor axes in order to be able to associate columns of the A matrix with specific source elemental concentration profiles. This target transformation rotation was first developed by Malinowski and coworkers (20,21). A suggested source profile is provided and a least-squares minimization is performed to rotate a factor axis toward this input test vector. Rewriting equation 19 yields

$$X = ARR^{-1}F \qquad\qquad\qquad (20)$$

where AR contains representations of the concentration profiles of the real sources and $R^{-1}F$ are the contributions of these sources to each sample. A rotation vector, r, a column of matrix R, is found by using a least-squares fit to a possible test vector b. The vector may be calculated by

$$r = (A^{t}WA)^{-1} A^{t}Wb \qquad\qquad\qquad (21)$$

where A^{t} is the transpose of A and W is a weighting matrix. The weighting matrix W is a diagonal matrix with the diagonal terms being the elemental weights to be used in this least-squares fit. The elemental weights that can be used are any that represent the statistical variation or confidence in the elemental data, e.g., the inverse of the square of the average experimental error, the variance of the elemental concentrations in the data set, or the error-weighted variance of the elemental concentrations. If there is to be no weighting, the diagonal terms are simply set equal to one. Details of the derivation of equation 2 is given by Malinowski and Howery (20).

While trying to resolve which sources are present in the data, one starts with an initial guess of the elemental composition of the source material. This concentration profile is then used as the test vector, b, in equation 21. From the rotation vector and b, a predicted vector, b′, can be calculated. The error observed between the original test vector b and the predicted test vector b′ gives an indication as to whether the test vector is a reasonable representation of a factor. Then b′ can be used as the new initial test vector b and a new predicted b″ can be calculated. Thus, the original b has been refined to a b that better represents the data. Continuing in this manner, one can iterate the initial guess of b toward a b′ that is much more representative of the specific sources for that particular data set.

One of the claims of factor analysis is that a minimum of prior knowledge is required, yet the target transformation rotation begins with an initial test vector. The question now arises, how good must the initial test vector be? Must the initial vector be a close approximation to the result b or can a simple initial guess for b be used? To answer this question, Roscoe and Hopke (22) made a comparative study on a previously source resolved set of geological data (23). They found that source composition profiles could be developed by this iterative process from simple initial test vectors that consisted of zero values for all but one element and unity for that single element. They obtained excellent agreement between the source profiles developed by the target transformation rotation (22) and those given by the

earlier report ($\underline{23}$) as shown in table 5. Thus, the input test vectors
supplied appear to speed the convergence but do not necessarily set the
final values. It should be noted that in many instances the ultimate
source profiles are deduced from the simple vectors that are unique for
a single element and not from other test vectors. It does not appear
that a priori knowledge of the detailed source vectors for a particular
study area are required but further studies of this question are in
progress. For any of the test vectors, the nature of the analysis is
such that the relative concentrations of the elements are predicted, but
the absolute concentrations are not. If the total mass of the aerosol
sample has been measured, it is then possible to determine a set of
scaling factors. The F values for a sample are relative measures of the
mass contribution of each source to that sample and, therefore, the F
values should sum to the total mass if they have been properly scaled.
The concentration of an element can be rewritten as

$$x_{ij} = \sum_{k=1}^{p} a_{ik} F_{kj} = \sum_{k=1}^{p} \frac{a_{ik}}{b_k} (b_k F_{kj}) \tag{22}$$

such that the mass of the jth sample is given by

$$M_j = \sum b_k F_{kj} \tag{23}$$

The collection of measured M_j values and the calculated F_{kj}
values can then be used in a multiple regression analysis to determine
the b_k values. This analysis provides several tests. First, no
concentration value, (a_{ik}/b_k), should be greater than 100%. The
sum of these values over the m elements should be <100%. The typical
regression programs available calculate a constant additive term that
should be insignificant if the analysis has produced the right results.

Results. For the St. Louis data, the target transformation analysis
results for the fine fraction without July 4th and 5th are given in
table 6. The presence of a motor vehicle source, a sulfur source, a
soil or flyash source, a titanium source, and a zinc source are
indicated. The sulfur, titanium and zinc factors were determined from
the simple initial test vectors for those elements. The concentration
of sulfur was not related to any other elements and represents a
secondary sulfate aerosol resulting from the conversion of primary
sulfur oxide emissions. Titanium was found to be associated with
sulfur, calcium, iron, and barium. Rheingrover ($\underline{24}$) identified the
source of titanium as a paint-pigment factory located to the south of
station 112. The zinc factor, associated with the elements chlorine,
potassium, iron and lead, is attributed to refuse incinerator emissions.
This factor could also represent particles from zinc and/or lead
smelters, though a high chlorine concentration is usually associated
with particles from refuse incinerators ($\underline{25}$). The sulfur concentration
in the refined sulfate factor is consistent with that of ammonium
sulfate. The calculated lead concentration in the motor vehicle factor
of ten percent and a lead to bromine ratio of about 0.28 are typical of
values reported in the literature ($\underline{26}$). The concentration of lead in

Table 5.

Element	Phase A				Phase B			
	Variance	Weighted Variance	Average Error	Reported	Variance	Weighted Variance	Average Error	Reported
Th	17.9	17.9	17.9	17.6	1.81	1.82	1.84	2.5
Sm	5.86	5.86	5.85	5.85	4.61	4.61	4.61	4.6
U	6.74	6.75	6.74	6.7	.54	.54	.55	.72
Na	27700.	27700.	27700.	27600.	24600.	24600.	24600.	24300.
Sc	4.47	4.45	4.47	4.76	39.0	39.0	38.9	37.5
Mn	122.	121.	122.	134.	1420.	1420.	1420.	1370.
Cs	16.1	16.1	16.1	15.9	0.	0.	0.	.50
La	23.0	23.0	23.0	22.7	12.5	12.5	12.5	12.5
Fe	6040.	6000.	6060.	6700.	80100.	80000.	80000.	77000.
Al	65100.	65100.	65000.	65000.	84300.	84300.	84300.	84000.
Dy	7.73	7.73	7.72	7.80	7.39	7.39	7.39	7.1
Hf	3.80	3.80	3.80	3.8	3.18	3.18	3.18	3.2
Ba	87.3	87.2	87.2	-- a	135.	135.	135.	-- a
Rb	251.	251.	250.	245.	5.26	5.44	5.66	27.
Ce	53.5	53.5	53.4	52.8	27.2	27.2	27.2	27.6
Lu	.52	.52	.52	.52	.42	.42	.42	.43
Nd	26.9	26.9	26.9	25.6	15.3	15.3	15.3	18.
Yb	4.09	4.09	4.08	4.0	2.69	2.69	2.69	2.7
Tb	1.06	1.06	1.06	1.05	.79	.80	.80	.81
Ta	1.18	1.18	1.18	1.18	.48	.48	.48	.52
Eu	.09	.09	.09	.10	1.64	1.63	1.63	1.6
K	46500.	46500.	46400.	46000.	2480.	2510.	2550.	8000.
Sb	1.13	1.13	1.13	1.1	.20	.20	.20	.3
Zn	42.8	42.7	42.8	45.	118.	118.	118.	114.
Cr	7.92	7.82	7.96	9.4	186.	186.	186.	178.
Ti	233.	228.	235.	300.	9380.	9370.	9360.	9000.
Co	0.	0.	0.	.40	41.6	41.6	41.6	39.6
Ca	9490.	9450.	9500.	10400.	78900.	78800.	78800.	75000.
V	6.89	6.72	6.96	9.	302.	301.	301.	300.

a - not reported by Bowman et al., 1973

Table 6. Refined Source Profiles. (mg/g)
RAPS Station 112, July and August 1976

Fine Fraction, July 4th and 5th Excluded

Element	Motor Vehicle	Sulfate	Flyash/ Soil	Paint	Refuse
Al	5.	1.1	53.	0.0	0.0
Si	0.0	1.9	130.	0.0	7.
S	0.02	240.	19.	6.	0.0
Cl	2.4	1.1	0.0	4.6	22.
K	1.4	1.6	15.	5.7	48.
Ca	11.	0.0	16.	34.	1.2
Ti	0.0	0.7	2.5	110.	0.0
Mn	0.0	0.0	0.7	4.8	8.6
Fe	0.0	1.1	36.	90.	36.
Ni	0.08	0.04	0.042	0.011	0.7
Cu	0.6	0.01	0.0	0.0	8.7
Zn	0.8	0.0	0.0	3.7	65.
Se	0.1	0.1	0.001	0.2	0.2
Br	30.	0.03	2.5	0.0	0.05
Sr	0.09	0.01	0.15	0.1	0.005
Ba	0.7	0.05	0.07	28.	0.5
Pb	107.	6.5	5.	0.0	46.

motor vehicle exhaust will vary from city to city and is dependent on
the ratio of leaded gasoline to unleaded and diesel-powered vehicles.
The calculated source profile for the refuse factor, with high
concentrations of chlorine, zinc, and lead, is similar to that measured
by Greenberg and coworkers(25) for the Nicosia Municipal Incinerator
near Chicago. However, Greenberg et al found that chlorine, zinc, and
lead concentrations of 27%, 11%, and 7%, respectively. In the present
study, the chlorine concentration is only 2% and the zinc and lead
concentrations are half those found by Greenberg. The lower calculated
concentrations may result from the combining of both refuse-incinerator
and lead/zinc-smelter emissions into a single factor. In the
paint-pigment component, the titanium and iron concentrations are
similar to those calculated by Dzubay (27). The nature of the
calculated soil/flyash factor is more like that of coal flyash than
soil, though the absolute concentrations of the major elements are less
than those reported by Gladney (28) and Fisher, et al. (29) for coal
flyash. Because of the similarity of their elemental profiles,
differentiating soil and coal flyash is a problem often encountered in
aerosol source resolutions. Coal flyash emissions are expected to
contribute more to the fine fraction while crustal material should be
found in the coarse fraction. Thus, we conclude that this factor is
primarily the result of coal-burning power plant emissions. Reliable
data for elements, such as arsenic, would be needed to clearly
differentiate the contributions of soil and coal flyash.

For the coarse fraction, the target transformation indicated the
presence of a sulfate factor, a paint-pigment factor, and two crustal
factors; an alumino-silicate type and a limestone or cement source. In
an earlier factor analytical study of aerosol sources in the St. Louis
area, Gatz (8) found the element calcium to be associated with other
than crustal sources. The high calcium loading at one site was
attributed to cement plants in the sampling area. Kowalczyk (30)
reports finding a potassium to calcium ratio of 0.8 in plume aerosols
collected over a cement plant near Washington, D. C. The uniqueness
test for calcium shows no strong correlation between calcium and
potassium, indicating the origins of the source are probably crustal.
The refined source profiles that best reproduced the coarse fraction are
listed in table 7. The calculated profiles of the two crustal
components follow those of Mason (31), though the calcium concentration
of 20% in the limestone factor is less than the reported value. The
paint pigment profile strongly resembles that calculated for the
fine-fraction data. The only major difference is that unlike the fine
fraction, the coarse-fraction profile does not associate barium with the
paint-pigment factor. The calculated sulfur concentration in the
coarse-fraction sulfate factor is much less than that in the
fine-fraction and there are sizable concentrations of elements such as
aluminum, iron, and lead not found in the fine-fraction profile. The
origin of this factor is not clear although as described earlier a
possible explanation is that a small part of the sulfate particles in
the fine fraction ended up in the coarse samples.

Table 8 summarizes the average elemental concentrations along with
the average observed concentrations for the fine fraction. The major
elements, Al, Si, S, K, Ca, Fe, and Pb are fit very closely. The
overall fit for the remaining elements is also fairly good. However,
the average point-by-point errors in the reproduced data range up to a
value of 350% for barium. Note that despite the large point-by-point

Table 7. Refined Source Profiles. (mg/g)
RAPS Station 112, July and August 1976

Coarse Fraction

Element	Soil	Limestone	Sulfate	Paint
Al	71.	30.	28.	5.
Si	274.	150.	0.0	0.0
S	4.9	0.0	90.	37.
Cl	1.4	16.	3.6	6.9
K	19.	15.	9.3	0.0
Ca	40.	188.	0.08	25.
Ti	0.0	1.5	0.0	128.
Mn	0.8	1.6	1.	1.2
Fe	40.	34.	43.	65.
Ni	0.01	0.2	0.2	0.06
Cu	0.0	0.6	0.9	0.0
Zn	0.0	4.2	4.3	0.3
Se	0.01	0.02	0.13	0.001
Br	0.5	3.1	3.8	1.3
Sr	0.2	0.3	0.2	0.1
Ba	0.6	0.7	0.4	3.2
Pb	1.5	11.	13.	6.7

Table 8. Summary of Mass Contributions. (ng/m^3)
RAPS Station 112, July and August 1976

Samples From July 4th and 5th Excluded

Fine Fraction

Element	Motor Vehicle	Sulfate	Flyash/Soil	Paint	Refuse	Total Predicted	Total Observed [a]	Average % Error [b]
Al	23.	21.	170.	0.0	0.0	230	200 ± 24	29
Si	0.0	36.	420.	0.0	9.	470	450 ± 59	15
S	0.09	4570.	62.	2.	0.0	4630	4360 ± 320	7
Cl	11.	21.	0.0	2.0	27.	61	80 ± 9	163
K	6.	30.	49.	2.4	60.	150	150 ± 9	28
Ca	49.	0.0	52.	14.	1.5	120	110 ± 10	53
Ti	0.0	12.	8.	46.	0.0	66	64 ± 13	260
Mn	0.0	0.0	2.3	2.0	11.	15	17 ± 3	122
Fe	0.0	21.	120.	37.	45.	220	220 ± 19	15
Ni	0.4	0.8	0.1	0.0	0.9	2.1	2.2 ± 0.2	106
Cu	2.6	0.2	0.0	0.0	11.	14	15 ± 2	210
Zn	3.6	0.0	0.0	1.5	81.	86	75 ± 8	77
Se	0.4	2.1	0.0	0.1	0.2	2.8	2.7 ± 0.2	89
Br	135.	0.5	8.2	0.0	0.1	140	132 ± 8	27
Sr	0.4	0.2	0.5	0.1	0.0	1.2	1.1 ± 0.1	88
Ba	3.1	0.9	0.2	12.	0.6	16	15 ± 4	353
Pb	480.	120.	16.	0.0	57.	680	720 ± 53	6

[a] Uncertainty is the standard deviation of the mean value.

[b] Average point-by-point error in predicted and measured data.

error, the average predicted and observed concentrations are very close, 16 and 15 ng/m respectively. The disparity is, in part, due to the tendency of the target transformation rotation to produce profiles that represent the average composition of each source. Thus, even though the average fit for barium is very good, the point-by-point error indicates that this element has not been well reproduced. The problem is compounded by the large number of values below detection limits for many of the less abundant elements. The presence of source components that are not resolved can lead to the underprediction of elements. Table 9 summarizes the mass contributions for the coarse fraction. Here, all the elements are fit well.

A comparison of the predicted and measured masses for each sample is another indicator of the quality of fit produced by the target transformation. The average deviations in the mass predictions were 16% for the fine-fraction data and 12% for the coarse. The very good fit to the mass predictions indicates that most of the undetermined elements such as carbon and nitrogen correlate fairly closely with the measured elements. Figure 3 summarizes the average mass contribution for each source to the total measured sample mass. The low percentage of unaccounted sample mass is expected in this type of analysis since the regression fit calculates scaling factors so as to minimize the overall difference between the measured and predicted sample mass. However, possible uncertainties in the scaling factors of the less important sources, i. e. refuse, paint-pigment, and flyash, could result in large uncertainties in the calculated concentrations of these sources. Secondary sulfate aerosol particles account for 64% of the mass of the fine-fraction data, an average concentration of about 19 g/m^3. Motor vehicle emissions account for another 15%. The measured lead concentration is divided among the refuse and motor vehicle factors. Here the lead contribution is 70% from motor vehicle emissions and 10% from refuse incinerators. In the coarse fraction, the two crustal components account for 80% of the total mass.

This approach has clearly allowed the resolution of the sources with results that appear to be very competitive to the chemical mass balance method. However, it was not necessary to make initial assumptions regarding the number of particle sources or their elemental composition. Additional studies need to be made to test the accuracy and precision with which such resolutions can be made.

Conclusions

It is clear that several forms of factor analysis can be very useful in the interpretation of aerosol composition data. The traditional forms of factor analysis that are widely available permit the identification of sources, the screening of data for noisy results, and the identification of interferences or analytical procedure problems. It is important, however, that new users of these techniques take the time to develop a little understanding of the strengths and limitations of them. It is very easy to employ a standard statistical package with

Table 9. Summary of Mass Contributions. (ng/m^3)
RAPS Station 112, July and August 1976

Coarse Fraction

Element	Soil	Limestone	Sulfate	Paint	Total Predicted	Total[a] Observed	Average[b] % Error
Al	1365.	272.	111.	125.	1760	1840 ± 180	12
Si	5266.	1359.	0.0	0.0	6630	6400 ± 580	4
S	94.	0.0	356.	92.	543	490 ± 32	35
Cl	27.	145.	14.	17.	200	210 ± 213	72
K	365.	136.	37.	0.0	540	540 ± 50	9
Ca	769.	1703.	0.3	62.	2540	2380 ± 141	8
Ti	0.0	14.	0.0	320.	330	300 ± 43	107
Mn	15.	15.	4.0	3.0	37	39 ± 4	37
Fe	769.	308.	168.	162.	1410	1470 ± 158	11
Ni	0.2	2.0	0.8	0.1	3.2	3.6 ± 0.4	155
Cu	0.0	5.8	3.6	0.0	9.4	8.9 ± 0.8	149
Zn	0.0	38.	17.	0.7	56	56 ± 6	126
Se	0.2	0.2	0.5	0.0	0.9	0.7 ± 0.04	154
Br	8.8	28.	15.	3.2	55	51 ± 3	51
Sr	3.8	2.4	0.7	0.3	7.2	7.9 ± 0.8	21
Ba	11.	6.3	1.4	8.0	27	27 ± 3	53
Pb	29.	100.	52.	17.	197	193 ± 11	47

[a] Uncertainty is the standard deviation of the mean value.

[b] Average point-by-point error in predicted and measured data.

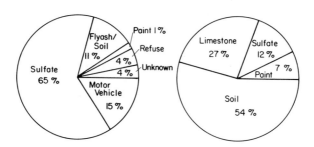

Figure 3. Average percent contribution of each source to the total average mass. The data are the fine- ((left) average mass = 29.4 µg/m³) and coarse-fraction ((right) average mass = 32.5 µg/m³) samples from RAPS Station 112 for July and August 1976.

standard options without understanding the explicit and implicit assumptions that have thus been made. A new form of factor analysis has been applied to aerosol source resolution. It requires that variables that are not linearly additive properties of the system be excluded. However, it is possible to identify the number of sources, their elemental composition and the amount of mass they contribute to the ambient aerosol. A major limitation to the method is the quality of data. Particularly if the regression approach is to be used to determine scaling factors, the total sample mass must be measured. Also the elemental analysis should be sufficiently complete to account for a relatively high fraction of the total mass observed if a factor analysis is to be performed such that control strategies could be based on its results. Although it appears that an excellent fit to the St. Louis data was obtained without the measurement of carbon or nitrogen, it would seem likely that there is a strong relationship between N and S as $(NH_4)_2SO_4$ and that most of the carbon in the summer months is strongly correlated with other elements such as lead. The results of Macias and Chu (32) demonstrate that there is a very strong correlation between elemental carbon and both lead and bromine in St. Louis. Large sources of uncorrelated carbon would lead to much poorer quality results. It is, therefore, important in the planning new air sampling programs to include the requirements of the endpoint statistical analysis so that the final source resolution will have validity.

Acknowledgements

    I would like to acknowledge the contributions of Daniel Alpert and Bradley Roscoe in the development and continuing exploration of factor analysis. This work has been supported in part by the University of Illinois Campus Research Board, the U. S. Environmental Protection Agency (Contracts D6004NAEX and 68-02-3449 and Grant R808229) and the U. S. Department of Energy (Contract DE-AC02-80EV10403.A000).

Literature Cited

1. Rozett, R. W.; Petersen, E. M. Methods of Factor Analysis of Mass Spectra, Anal. Chem., 1975, 47, 1301.

2. Rozett, R. W.; Petersen, E. M. Classification of Compounds by the Factor Analysis of their Mass Spectra, Anal. Chem., 1976, 48, 817.

3. Duewer, D. L.; Kowalski, B. R.; Fasching, J. L. Improving the
    Reliability of Factor Analysis of Chemical Data by Utilizing
    the Measured Analytical Uncertainty, Anal. Chem., 1976, 48,
    2002.

4. Hopke, P. K.; Gladney, E. S.; Gordon, G. E.; Zoller, W. H.;
    Jones, A. G. The Use of Multivariate Analysis to Identify
    Sources of Selected Elements in the Boston Urban Aerosol,
    Atmospheric Environ., 1976, 10, 1015.

5. Cattell, R. B. "Handbook of Multivariate Experimental
    Psychology", Rand McNally: Chicago, 1966; 174.

6. Blifford, Jr., I. H.; Meeker, G. O. A Factor Analysis Model of
    Large Scale Pollution, Atmospheric Environ., 1967 1, 147.

7. Gaarenstroom, P. D.; Perone, S. P.; Moyers, J. L. Application
    of Pattern Recognition and Factor Analysis for Characterization
    of Atmospheric Particulate Composition in Southwest Desert
    Atmosphere, Environ. Sci. Technol., 1977, 11, 795.

8. Gatz, D. F. Identification of Aerosol Sources in the St. Louis
    Area Using Factor Analysis, J. Appl. Met., 1978, 17, 600.

9. Sievering, H.; Dave, M.; Dolske, D.; McCoy, P. Trace Element
    Concentrations over Mid-Lake Michigan as a Function of
    Meteorology and Source Region, Atmospheric Environ., 1980, 14,
    39.

10. Lewis, C. W.; Macias, E. S. Composition of Size-Fractionated
    Aerosol in Charleston, West Virginia, Atmospheric Environ.,
    1980, 14, 185.

11. Kaiser, H. F. Computer Program for Varimax Rotation in Factor
    Analysis, Educational and Psychological Measurement, 1959, 19,
    413.

12. Hoffman G. L.; Duce, R. A. Copper Contamination of Atmospheric
    Particulate Samples Collected with Gelman Hurricane Samples,
    Environ. Sci. Technol., 1971, 5, 1134.

13. Changnon, S. A.; Huff, R. A.; Schickedenz, P. T.; Vogel, J. L.
    Summary of METROMEX, Volume 1: Weather Anomalies and Impacts,
    Illinois State Water Survey Bulletin 62, Urbana, IL, 1977.

14. Ackerman, B., et al.. Summary of METROMEX, Volume 2: Causes of
    Precipitation Anomalies, Illinois State Water Survey Bulletin
    63, Urbana, IL, 1978.

15. Kaiser, H. F.; Hunka, S. Some Empirical Results with Guttmans
    Stronger Lower Bound for the Number of Common Factors,
    Educational and Psychological Measurement, 1973, 33, 99.

16. Imbrie, J.; Van Andel, T. H.   Vector Analysis of Heavy-Mineral Data, Geological Soc. Amer. Bull., 1964, 75, 1131.

17. Miesch, A. T.   Q-Mode Factor Analysis of Geochemical and Petrologic Data Matrices with Constant Row-Sums, U. S. Geological Survey Professional Paper 574-G, Washington, 1976.

18. Alpert D. J.; Hopke, P. K.   A Quantitative Determination of Sources in the Boston Urban Aerosol, Atmospheric Environ., 1980, 14,   .

19. Alpert, D. J.; Hopke, P. K.   A Determination of the Sources of Airborne Particles Collected During the Regional Air Pollution Study, Atmospheric Environ., in press, 1981.

20. Malinowski E. R.; Howery, D. G.   "Factor Analysis in Chemistry", John Wiley & Sons, Inc.: New York, 1980.

21. Weiner, P. H.; Malinowski, E. R.; Levinstone, A. R.   Factor Analysis of Solvent Shifts in Proton Magnetic Resonance, J. Phys. Chem., 1970, 74, 4537.

22. Roscoe, B. A.; Hopke, P. K.   Comparison of Weighted and Unweighted Target Transformation Rotations in Factor Analysis, Computers and Chemistry, in press.

23. Bowman, H. R.; Asaro, F.; Perlman, I.   On the Uniformity of Composition in Obsidians and Evidence for Magnetic Mixing, J. Geology, 1973, 81, 312.

24. Rheingrover, S. W.   A Statistical Model for Titanium Pollution Transport and Dispersion in the Atmosphere of Saint Louis, M.S. Thesis, Florida State University, 1977.

25. Greenberg, R. R,; Gordon, G. E.; Zoller, W. H.   Composition of Particles from the Nicosia Municipal Incinerator, Environ. Sci. Technol., 1978, 12, 1329.

26. Dzubay, T. G.; Stevens, R. K.; Richards, L. W.   Composition of Aerosols over Los Angeles Freeways, Atmospheric Environ, 1979, 13, 653.

27. Dzubay, T. G.   Chemical Element Balance Method Applied to Dichotomous Sampler Data, Annals N. Y. Acad. Sci., 1980, 338, 126.

28. Gladney, E. S.   Trace Elemental Emissions from Coal-Fired Power Plants: A Study of the Chalk Point Electric Generating Station, Ph.D Thesis, University of Maryland, 1974.

29. Fisher, G. L.; Crisp, C. E.; Hays, T. L.   Carbonaceous Particles in Coal Fly Ash. In Proceedings of the Conference on Carbonaceous Particles in the Atmosphere, Lawrence Berkeley Laboratory Report LBL-9037, CONF-7803131, UC-11, 1978.

30. Kowalczyk, G. S. Concentration and Sources of Elements on Washington, D. C. Atmospheric Particles, Ph.D Thesis, University of Maryland, 1979.

31. Mason, B. "Principles of Geochemistry"; John Wiley & Sons, Inc.: New York, 1966.

32. Macias, E.S.; Chu, L.S., Urban Aerosol Carbon - Primary or Secondary?, in Chemical Composition of the Atmospheric Aerosol: Source/Air Quality Relationships, E.S. Macias and P.K. Hopke, Eds., A.C.S. Symposium Series, 1981.

RECEIVED April 10, 1981.

Composition of Source Components Needed for Aerosol Receptor Models

GLEN E. GORDON, WILLIAM H. ZOLLER, GREGORY S. KOWALCZYK[1], and SCOTT W. RHEINGROVER

Department of Chemistry, University of Maryland, College Park, MD 20742

A critical requirement for the success of receptor models for atmospheric particles is that the
compositions of particles from all major sources in
an area be accurately known. Chemical element balances (CEBs) of 130 samples taken in Washington,
D.C. and analyzed for 40 elements yielded nearly the
same source strengths when 28 elements are used in
the least-squares fit as when only nine carefully
chosen elements are used. Certain elements are important to the stability of CEB fits (Na, Ca, V, Mn,
As and Pb) and should be measured carefully in particles from sources. For three of the nine elements
(Al, Fe and Zn), other elements can serve as surrogates (many lithophiles for Al and Fe, Sb and Cd for
Zn). Measurements on many more sources of each important type should be done in order that trends can
be observed that will allow one to predict compositions of particles from unmeasured sources. In-
stack measurements should include collections of at
least two size fractions of particles plus vapor-
phase species. Measurements of at least 20 elements
plus some classes of carbonaceous material should be
made.

Because of the uncertainties in the use of source-emissions
inventories to estimate contributions from various sources to ambient levels of suspended particles, many workers have been developing and testing aerosol receptor models (1). The basic idea of
receptor models is that chemical compositions of particles from
various types of sources are sufficiently different that one can
determine contributions from the sources by making detailed measurements of the compositions of ambient aerosols and of particles
from the sources. Several computational methods have been used

[1]Current address: Northeast Utilities, Hartford, CT 06101

in the general framework of receptor models, including chemical
element balances (CEBs) (2), factor analysis (3) and target-
transformation factor analysis (4), but here we examine the general needs for measurements required for all receptor-model interpretations, using the CEB approach below to test the sensitivity
of the results to the use of various measured elements. Specifically, we focus on the question of sources: what types of information are most needed about particles from various sources for input to receptor models? In particular, which elements or other
chemical species provide the most nearly unique "fingerprints" of
the important class of sources?

These questions were a major theme of the NSF/RANN-sponsored
project conducted at Maryland from 1972 to 1980. Much of the
group's work was directed towards the question: Are compositions
and size distributions of particles from different sources sufficiently distinct that one can detect their components by analysis
of ambient particles? If so, which elements yield the clearest
distinctions? To answer these questions, particles from many
sources were investigated: two coal-fired power plants (5, 6, 7),
an oil-fired plant (8), three municipal incinerators (9, 10), five
copper smelters (11, 12), a lime kiln and a steel mill (13), motor
vehicles (14), and continental background aerosols and local soil
(15). To test receptor models, Kowalczyk (16) collected 130
whole-filter samples from ten sites in the Washington, D.C. area,
augmented with size information from cascade impactors. Washington is a good test area for receptor models, as it contains little
industrial activity. Thus, if receptor models are to work at all,
they should work in Washington. It should be possible to account
for ambient concentrations of most elements with linear combina-
tions of concentration patterns of components for a few common
types of sources from among those measured (soil, coal, oil, motor
vehicles, incinerators) and other common aerosol material (sea
salt, limestone).

Before discussing the results, we should note several philo-
sophical points regarding the design of these types of experiments:

First, we analyzed samples for a large number of elements to
identify any elements, regardless of toxicity or typical concentration, that would provide signals for the presence of material
from certain types of sources. Both ambient samples and particles
from sources were analyzed by instrumental neutron activation
analysis (INAA), by which one can often measure about 35 elements
in individual samples (17). As the important elements Pb, Ni and
Cd are not consistently, if ever, observed by INAA, they were often measured by other methods. As INAA is sensitive to very small
amounts of obscure elements, we have obtained reliable data for
elements such as Ga, Hf, Sc, In, W and many rare earths which pose
no known health hazard at present levels and contribute insignificant amounts of mass to TSP. However, as discussed below, many
trace elements have already been shown to be important in receptor

applications (e.g., As, Sb, Mn, Cd) and others may be useful once we understand sources better.

Second, because of the high sensitivity of INAA, we collected such small masses of material that samples could not be conveniently weighed. Thus, most results were obtained as the mass of an element (on particles) per unit volume of air or stack gas rather than concentration in the suspended matter. Because of this, we can compute mass contributions from the sources to ambient air only with independent knowledge of the concentration of at least one key element in particulate matter from each source; e.g., of V in particulate matter from oil-fired power plants. It would be desirable to have the information on mass concentrations; however, the point of our study was the development and testing of the fundamentals of receptor models and, only incidentally, to estimate TSP contributions from various sources.

Third, in studying sources, whenever practical we analyzed in-plant materials (e.g., coal, oil, various ashes) as well as suspended materials released to the atmosphere. By studying the fractionation of elements upon passage through plants, we hoped to obtain results that would be transferable to other similar sources. For example, it would be desirable to apply receptor models to a city having many coal-fired plants without having to collect and analyze particles from each plant. If we knew the fractionation of elements between the gas and particulate phases and among particles of different sizes as a function of temperature, etc., we could perhaps predict compositions and size distributions of particles from a particular plant by knowing the composition of coal, the type of boiler and the efficiency curves of the pollution control devices. Although physical chemical models of the behavior of elements in coal-fired plants exist (e.g., Ref. 18), there haven't been enough careful studies of different coal-fired plants to determine whether or not the results are accurately transferable (discussed below).

Chemical Element Balances

In view of our knowledge of the composition of particles from most sources in the Washington area, it is appropriate to use the CEB method rather than methods that extract source compositions from data on ambient particles. According to the CEB method, the concentration of an element i in a receptor sample is given by

$$C_i = \Sigma m_j x_{ij} \tag{1}$$

where m_j is the mass of material contributed by source j and x_{ij} is the concentration of element i particles from source j. Given the compositions of the important components, i.e., the x_{ij}s, the object is to find the source strengths, i.e., the m_js, that give a good fit to the observed concentrations of the elements at the receptor, the C_is. Usually this is done by a least-squares fit to observed concentrations of selected elements.

Differing philosophies are used in the selection of elements to be fitted. Ours has been that, as we are testing the CEB approach, we should select a minimum of "marker" elements (i.e., elements whose concentrations are quite sensitive to amounts of specific sources) to be fitted in order to leave many "floating" elements for testing the fit. Also, looking forward to eventual routine use of receptor models, one would like to specify a minimum of measurements needed to determine source contributions to TSP. Thus, we have often used eight carefully selected marker elements to determine strengths of six components (19) or nine marker elements for seven sources (20). Five markers are quite obvious: Na for sea salt, V for oil, Ca for limestone, Zn for refuse combustion and Pb for motor-vehicle emissions. A major problem for all receptor models is that of distinguishing between soil and emissions from coal combustion, which are quite similar for many elements. We use Al and Fe as measures of the sum of these components and, based on the source studies noted above, As as a measure of the coal component and Mn for soil.

In Table I, are shown results of one of our best CEBs for Washington-area aerosols. This CEB was done using 28 marker elements, but the results are almost indistinguishable from those obtained with nine markers (see below). The quantity L/S, the "larger-over-smaller" ratio, is a measure of the quality of the fit. For the CEB of each sample, the L/S value is the Predicted/Observed ratio or the inverse, whichever is larger, so the value is always unity or greater. The L/S values listed in Table I are averages of the values for each sample. They are a better measure of the quality of the fit than Predicted/Observed ratios, as the average of the latter may be close to unity because of compensation of larger and smaller values, but L/S reflects this fluctuation about perfect fits. For example, for Sc and Mn in Table I, Pred./Obs. is about 1.0, but L/S is 1.2, indicating ±20% fluctuations about perfect fits. By contrast, Na has values of unity for both quantities, meaning that Na is fitted perfectly by each CEB.

We do not include L/S values of volatile elements (halogens, Se, W) in the overall average, as we don't expect to fit those non-conservative elements well. The average value for the remaining 35 elements is 1.94. The CEB provides a surprisingly good fit to several elements with very low concentrations, e.g., Sc, Rb, Sr, Ga, Ag, Cd, Th and light rare earths. Most elements are reasonably well fitted, with L/S values <1.5, but a few are poorly fitted and raise the average considerably. We are not too concerned about poor fits for elements such as In and Cs, as their concentrations in the source materials are not well known. The poor fits for K, Mg, Cr, Cu and Ni are more serious, but we do not have components in the CEB with high enough ratios of K and Mg to Ca and of Cr, Cu and Ni to other elements to account for these

Table I. Chemical Element Balances for 130 Washington, D.C. Ambient-Particle Samples from Ten Sites (Aug/Sept, '76)[a]

Element	Average L/S[b]	Pred/Obs	Major Source(s)
Na[c]	1.00[d]	1.00	Marine
Mg[c]	1.62[d]	0.62	Limestone
Al[c]	1.16[d]	1.00	Soil, Coal
Cl	9.2	5.8	Marine
K[c]	1.74[d]	0.61	Soil
Ca[c]	1.09[d]	0.95	Limestone
Sc[c]	1.20[d]	1.00	Coal, Soil
Ti[c]	1.34[d]	0.77	Soil, Coal
V[c]	1.08[d]	1.06	Oil
Cr	6.9[d]	0.14	Coal, Soil
Mn[c]	1.27[d]	1.01	Soil
Fe[c]	1.18[d]	0.92	Soil, Coal
Co[c]	1.25[d]	0.78	Coal, Soil
Ni	4.4[d]	0.35	Oil
Cu	3.0[d]	0.36	Coal, Motor Vehicle
Zn[c]	1.37[d]	0.78	Refuse
Ga[c]	1.9[d]	0.66	Coal, Soil
As[c]	1.55[d]	1.05	Coal
Se	4.1	0.36	Coal
Br	1.72	1.23	Motor Vehicle
Rb[c]	1.54[d]	0.83	Soil, Coal
Sr[c]	1.40[d]	0.86	Soil, Coal
Ag[c]	1.48[d]	1.25	Refuse
Cd[c]	1.81[d]	0.75	Auto, Refuse
In	4.9[d]	0.31	Refuse, Coal
Sb[c]	1.55[d]	0.69	Refuse
I	2.4	1.73	Coal
Cs[c]	3.3[d]	0.42	Coal, Soil
Ba	1.38[d]	1.05	Soil, Motor Vehicle, Coal
La[c]	1.42[d]	0.73	Soil
Ce[c]	1.30[d]	0.81	Soil, Coal
Sm[c]	1.57[d]	0.66	Soil, Coal
Eu[c]	1.33[d]	1.02	Soil, Coal
Yb[c]	2.6[d]	2.1	Soil, Coal
Lu	3.4[d]	2.8	Coal, Soil
Hf[c]	1.8[d]	0.55	Soil, Coal
Ta[c]	1.8[d]	1.4	Soil
W	4.6	0.25	Coal
Pb[c]	1.14[d]	1.06	Motor Vehicle
Th	1.19[d]	0.90	Soil, Coal
Avg.	1.94		

[a]CEBs used 28 marker elements, ignoring three points each for Ta and Eu. The Mn concentration was reduced by factor of 0.7 (see text).
[b]Larger/Smaller ratio = Pred/Obs or Obs/Pred, whichever is larger.
[c]Marker element. [d]Included in overall avg. L/S.

elements. Perhaps some components do not have high enough concentrations of these elements or, more likely, some sources of these
elements may not be included in the CEB (20). Manganese presents
a particular problem: although it is important for fitting the
soil component (see below) and the CEB predicts that most of the
Mn is associated with soil, measurements of the size distribution
of Mn-bearing particles show that about 30% of the Mn is attached
to fine particles from some other source (20). Thus, for most
CEBs, we have reduced the observed Mn concentrations by a factor
of 0.7 before attempting the fits.

Choice of Marker Elements. In contrast to our philosophy of
using a minimum of marker elements, others feel that one should
take full advantage of the information contained in the data by
using nearly all non-volatile elements as markers (21, 22). Table
II shows summaries of CEBs performed with various choices of marker elements. Note that, if a component is included that is not
present in some samples, one obtains negative source strengths for
that component resulting from fluctuations or errors in contributions from other components. Our program does not prevent the occurrence of negative values, but in most reasonable fits they occur quite rarely. For example, in the nine-marker fit in Table
II, out of the 910 source strengths (7 components for 130
samples), only nine are negative. Most are for the marine component on days when the wind was never from an easterly direction, so
the marine component was surely small.

The first CEB summary in Table II results from the use of the
nine marker elements for seven components. The fit is quite good,
with an average L/S of 1.93, but with nine negative source
strengths. The next summary is for 21 elements obtained by adding
twelve markers that are fairly well-behaved (see Table I) to the
original nine. The fit is almost the same as for the nine-marker
case. Since most added markers are lithophile elements, there is
a slight change in the average strengths of the "crustal" components, with an increase in the soil component and decrease in the
limestone.

The next CEBs included 26 marker elements by adding five
trace elements, some of whose concentrations were not well fitted.
Although there are slight changes in strengths of other components, the major change is a 25% reduction of the refuse component,
caused by the inclusion of Cd and Sb as markers. The marine
source strength increased slightly (because of decreased Na from
the refuse source) and most of the negative marine components became positive.

In the 28-marker set, we added Eu and Ta to the 26 from
above. Three datum points for each element appeared to be incorrect, but careful examination of the raw data yielded no justifi-
cation for changing or eliminating them. However, as they would
have greatly distorted CEBs for their samples, we had the fitting
routine ignore them. The resulting CEBs, shown in detail in Table

Table II. Summary of Chemical Element Balances for Washington, D.C. for Different Numbers of Marker Elements[a]

Marker Elements	Avg. L/Sb	Negative Source Strengths	TSP Contributions (µg/m³)						
			Soil	Marine	Coal	Oil	Refuse	Motor Vehicle	Limestone
9 (Na, Al, Ca, V, Mn, Fe, Zn, As, Pb)	1.93	9	14.1	0.60	3.8	0.32	0.83	4.0	1.9
21 (9 above plus Mg, K, Sc, Ti, Co, Sr, Cs, La, Ce, Sm, Hf, Th)	1.93	8	16.0	0.59	3.8	0.34	0.82	4.0	1.7
26 (21 above plus Ga, Rb, Ag, Cd, Sb)	1.95	2	15.6	0.63	3.9	0.34	0.62	4.2	1.7
28 (26 above plus Eu[c], Ta[c])	1.94	2	14.4	0.64	4.0	0.34	0.62	4.3	1.8
30 (28 above plus Br, Ba)	1.98	7	14.0	0.64	4.0	0.34	0.64	3.2	1.9
29 (26 above plus Cr, Ni, Cu)	1.94	2	15.6	0.62	4.0	0.35	0.64	4.2	1.7
20 (29 above minus group of 9 in Table I)	3.4	107	19.9	7.25	3.1	0.58	0.65	2.8	3.5

[a]Same data set as for Table I. Mn reduced by 0.7. [b]Larger/Smaller ratio = Pred/Obs or Obs/Pred, whichever is larger.

[c]Three points each for Eu and Ta ignored by CEB.

I, yielded results almost identical with those of the 26-marker
set. All things considered (including the number of negative
source strengths), this is probably the highest quality fit obtainable with the seven components used.

The 28 markers include a majority of the measured elements,
the remaining ones being Ba, Cr, Cu, Ni, In, Yb and Lu and the
volatile species Cl, Br, I, Se and W. We were concerned because
the strength of the motor-vehicle component is determined largely
by the Pb concentration and, thus, is subject to errors in the Pb
determination or to contributions of Pb from sources not included
in the CEB. To fix the component more reliably, we would prefer
to have additional markers that influence the component's
strength. We attempted to include Ba and Br as shown in the summary for thirty markers. Although Br is volatile, its concentration is reasonably well fitted by the previous sets of CEBs (see
Table I). However, the inclusion of these elements appreciably
increases L/S and the number of negative source strengths, and decreases the strength of the motor-vehicle component. The Pb concentration is seriously underpredicted in many individual CEBs,
yielding an overall Pred/Obs value of 0.82. Based on these results, we feel it is better not to include Br and Ba in the CEBs.

The Ba of the motor-vehicle component appears to come mainly
from diesel trucks, in part because of its use as a smoke suppres-
sant (14). Thus, the Ba associated with the component may be
quite variable with time and traffic mix. If Br were included,
its relative concentration in the component should probably be reduced by about 10% to account for losses of this volatile element
between the time of release and its collection at a receptor site.
These elements could probably be used as markers if the fitting
procedure had provision for weighting based on uncertainties of
relative concentrations of elements in the components (21). For
the present tests, we have not included this feature, but have included weighting factors of $1/\sigma^2$ for each element in the samples,
where σ includes uncertainties of the analytical measurements and
filter blank values. If the additional weighting were included,
the importance of Ba and Br in determining the strength of the
motor-vehicle component would be reduced relative to that of Pb
because of their greater uncertainties.

At the bottom of Table II, we show the effects of adding Cr,
Ni and Cu as markers to the 26-marker set. Surprisingly, the inclusion of these poorly fitted elements causes a negligible change
in the quality of the fits or the strengths of the sources. Apparently, once the fit includes the initial 26 elements, it is
sufficiently stable that it is not much affected by the addition
of three more, even though the latter are poorly fitted elements.
In part the lack of effect is due to the weighting factors $1/\sigma^2$,
which are relatively large for Cu and Ni; however, Cr has a large
weighting, as its concentration can be quite accurately determined
by INAA.

At the New York Academy of Sciences meeting in 1979, J. W. Winchester asked what would happen if the original nine markers were entirely replaced by other choices (23). The last set of CEBs in Table II was performed by removing them. The result is a very poor fit, with L/S increasing to 3.4 and the number of negative source strengths to 107. The average source strengths change dramatically, with the marine component increasing more than tenfold and that of limestone doubling. Clearly, some of the original nine markers are quite important for the CEBs!

In Table III, we examine the importance of each marker. The first or "standard" calculation is the 28-marker set of Tables I and II. Below that are shown the results obtained by removing each markers, one at a time. We see an enormous increase in the strength of the marine component when Na is removed. When it is absent, the strength increases, apparently in trying to fit minor sea-water elements such as K and Mg. (Recall that Cl and Br are not markers in the set of 28.) The removal of V and Pb causes the strengths of the oil and motor-vehicle components, respectively, to double. The removal of Ca has a slight effect on the limestone/soil ratio and the deletion of As and Mn has small, but non-negligible effects on the soil-coal ratio. The removal of Al and Fe has little effect because of the presence of many other lithophile elements among the remaining 27.

The results of this exercise are rather encouraging. Perhaps most important, the CEBs have a great deal of stability: if several key elements are included as markers, one obtains nearly the same result for fits ranging from those with nine markers to those with 28. Thus, in the argument between our philosophy of minimum markers and Watson's, of using as many as possible, these results say that it doesn't make much difference - at least for the seven components used in fitting the Washington, D.C. aerosol compositions. Even the addition of three poorly fitted elements, Cr, Cu and Ni, doesn't cause large changes, if many other markers give stability to the fits. However, there is little reason for including them when it is clear no component has high enough ratios of these elements to other markers to account for those three.

The results in Tables II and III show that certain elements are vital for obtaining strengths of some sources. The most important is Na for areas that have an appreciable marine component and probably also for areas where roads are salted. Lead, Ca and V are important for motor vehicle, limestone and oil components. Arsenic and Mn are fairly important for distinguishing between coal and soil, but the CEBs do not fluctuate wildly without them if many other markers are used. Many lithophile elements can serve as surrogates for Al and Fe, and Sb and Cd can serve in place of Zn for setting the refuse component. Other elements may also be important in areas that have more industrial activities than Washington, D.C. Some components are quite "fragile", in that they are highly sensitive to the concentrations of one or two

Table III. Effects on Chemical Element Balances of Washington, D.C. When Key Marker Elements are Removed Individually[a]

Marker Elements	Avg. L/Sb	Negative Source Strengths	TSP Contributions (μg/m³)						
			Soil	Marine	Coal	Oil	Refuse	Motor Vehicle	Limestone
28 (See Table I)	1.94	2	14.4	0.64	4.0	0.34	0.62	4.3	1.8
Less Na	2.12	4	14.5	7.15	4.0	0.34	0.62	4.2	1.6
Less Al	1.94	2	14.7	0.64	4.0	0.34	0.62	4.3	1.8
Less Ca	2.03	30	15.5	0.63	3.8	0.34	0.62	4.3	2.1
Less V	1.95	4	14.4	0.60	3.8	0.75	0.60	4.3	1.8
Less Mn	1.94	3	16.2	0.63	3.6	0.34	0.62	4.2	1.9
Less Fe	1.95	1	14.1	0.64	4.1	0.34	0.62	4.2	1.8
Less Zn	1.98	3	14.4	0.65	4.0	0.34	0.60	4.3	1.8
Less As	1.97	9	13.0	0.65	4.9	0.34	0.61	4.2	1.9
Less Pb	1.99	9	14.4	0.66	4.0	0.35	0.54	9.9	1.7

[a]Same data set as for Table I. Mn reduced by 0.7.

[b]Larger/Smaller ratio = Pred/Obs or Obs/Pred, whichever is larger.

two elements, especially the motor-vehicle (Pb) and marine (Na) components. Other elements or compounds that would provide additional determinations of the strengths of these sources are needed (see below).

Despite the stability of the CEBs, one cannot choose marker elements indiscriminantly. We have seen that the addition of Br and Ba to the 28-marker set causes instabilities of the motor vehicle component and the elimination of key markers causes various instabilities.

Implications for Source Studies. The results discussed above indicate the need for measurements of a number of elements in studies of particles both from sources and in ambient air. Measurements of Na, Pb, Ca, As, Mn and V are very important for use in receptor models. Aluminum and Fe are quite useful, but not essential if many other lithophile elements are measured (e.g., Si, Ti, Sc). Likewise, Zn is useful, but could be replaced by elements such as Sb and Cd. However, Al, Fe and Zn can usually be measured more easily than their surrogates. Iron and elements such as Cr, Mn, Co and Ni will be important in areas that have iron and steel industries and elements such as Cu, Zn, Pb and other chalcophiles in areas that have non-ferrous metal industries. Sources of the unexplained Mn, Cr, Cu, Ni, K and Mg in Washington need to be identified. Thus, we have a set of about 15 elements that should be measured as a minimum, plus others that may provide additional useful information, e.g., Br, Ba, Cd, Sb. Note that it is not sufficient to measure a given element only in particles from the dominant source of the element. For example, motor vehicles are the major source of Pb in most areas; however, significant amounts are released by refuse incinerators and nonferrous smelters. Thus, if the CEB is to determine the correct source strength for the motor-vehicle component, Pb contributions from the other, less important sources must be known.

Most major studies of sources and ambient air are done using techniques that measure many elements in individual samples, especially INAA and x-ray fluorescence (XRF). However, considering the elements that should be measured, neither technique is sufficient by itself. Most important elements can be measured by INAA, but with the important exceptions of Pb, Ni and, sometimes, Cd. Likewise, XRF cannot be used to measure Na and has considerable difficulty with trace elements such as As, V, Cd and Co unless they are present in unusually large amounts. Thus, both techniques must be augmented by other methods. We usually augment INAA with atomic absorption spectrometry for Pb, Ni and Cd. The Oregon Graduate Center group (21, 22) uses a combination of XRF and short-irradiation INAA. This combination of instrumental methods leaves the samples intact and covers most key elements, the major exception being As. The latter is not too important in the Portland area, as little, if any coal is burned there. If needed, As

could be determined by INAA by performing a second irradiation of
each sample for somewhat longer times.

Additional markers are needed for several important sources.
Sodium is the only adequate marker for the marine component (and
road salt) and it is not obvious that others can be developed.
Lead is the only strong marker for the motor-vehicle component,
but it is a direct measure only of emissions from vehicles that
use leaded gasoline. As vehicles equipped with catalytic convertors take over an increasing fraction of the fleet, the mass/Pb
ratio of suspended particles from motor vehicles increases. Eventually the Pb marker will disappear completely. Even now, the
ratio depends on the mix of diesel trucks and cars, cars and light
trucks burning leaded gasoline and those equipped with catalytic
convertors in the area under study. To obtain more definitive
CEBs one should have separate markers for the different groups of
motor vehicles. Useful markers will probably not be found among
the elements per se, as emissions from diesel and catalytic-
convertor-equipped engines are largely carbonaceous materials.
Therefore, markers based on carbonaceous species should be developed, e.g., abundance patterns of certain classes of organic
compounds such as polynuclear aromatic hydrocarbons or thermal de-
composition patterns. Fortunately, considerable progress is being
made in developing these markers (24, 25, 26). They should also
be of value for other sources of carbonaceous particles such as
wood stoves and fireplaces, home oil or gas furnaces and water
heaters, etc.

Source Measurements Needed

Here we focus on the types of sources that should be investigated, the strategies for sampling and the interpretation of results. To illustrate several problems, we consider coal-fired
power plants, which have been studied in more detail than other
sources.

Coal-Fired Power Plants. Detailed studies of compositions of
particles collected from stacks of about ten coal-fired power
plants have been performed (5, 6, 7, 27-36). In a few cases, the
composition has been determined for particles in several size
groups. Some measurements on suspended particles have been augmented by analyses of input coal and ash fractions.

As noted above, we had hoped that one could perform measurements on several sources of a particular class and observe trends
that would allow one reliably to predict compositions of particles
from other members of that class to avoid the need to study each
individual major source in an area before performing receptor-
model interpretations of ambient aerosols of that area.

One useful method for systematizing data on sources and ambient aerosols is the use of enrichment factors with respect to
the Earth's crust, EF_{crust}, defined by (37, 38):

$$EF_{crust} = (C_x/C_{Al})_{particles}/(C_x/C_{Al})_{crust} \qquad (2)$$

where $\underline{C}$s are concentrations of element X and Al in the particles
and in average crustal material (39). Aluminum is not unique, but
it is a major lithophile element that can be measured by either
INAA or XRF.   Several other lithophiles could be used as well in
areas that have no special sources of the element, e.g., Si, Ti,
Sc.   To reduce the dependence of $\underline{EF}$ values upon errors in Al measurements, in most cases below, we have renormalized the $\underline{EF}$ values
so that the average of values for five lithophile elements is
unity.
     The use of $\underline{EF}$ values allows us to set limits on possible
sources of elements.   In Figure 1, $\underline{EF}$ values for six cities are
compared with the ranges for particles from nine coal-fired power
plants. For lithophile elements such as Si, Ti, Th, K, Mg, Fe and
many others not shown, $\underline{EF}$ values are close to unity as expected,
as these elements have mainly crustal sources, i.e., entrained
soil and the aluminosilicate portion of emissions from coal combustion (see Table I).   Many other elements are strongly enriched
in some or all cities, and, to account for them, we must find
sources whose particles have large $\underline{EF}$ values for those elements.
Some are fairly obvious from the above discussions: Pb from motor
vehicles, Na from sea salt in coastal cities, and V and, possibly,
Ni from oil in cities where residual oil is used in large amounts
(Boston, Portland, Washington).
     Coal-fired power plants release very large amounts of parti-
culate material.   The question is, however, what fractions of the
various elements in ambient air can be accounted for by particles
from coal-fired plants?   A major fraction of an element can be
contributed by coal combustion only if (1) coal accounts for an
appreciable fraction of the Al in the local atmosphere and (2) the
$\underline{EF}$ value of the element on particles from coal combustion is as
great as for ambient particles. Only for those elements in Figure
1 for which there is considerable overlap between the $\underline{EF}$ ranges
for cities and for coal-fired plants can coal possibly be a major
contributor.   Even if there is overlap, coal is not necessarily a
major source, as condition 1 above may not be met. On this basis,
coal combustion could be a major source of many lithophiles plus
Cr, Ni, As, Se and, in cities where little residual oil is used
(Charleston and St. Louis), V. The very high $\underline{EF}$ values for As and
Se and low values for V and Ni in Charleston, where little oil and
a great deal of coal are burned, lends credence to this interpretation.
     Enrichments of several elements on particles from coal com-
bustion are too low for coal to be a major source: Na, K, Mn, Cu,
Zn, Cd, Sb and Pb.   Except possibly for Cu, whose major source is
unknown, these results agree with the CEBs for Washington, which
indicate other principal sources for those elements.   A possible
weakness in the argument presented is that it contains the implicit assumption that there is no fractionation of particles bearing

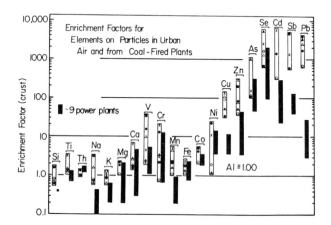

Figure 1.   Enrichment factors with respect to crustal abundances (39) for elements attached to urban aerosols from (●) Washington, D.C. (16), (○) Tucson, AZ (40), (✕) St. Louis, MO (based on data from Loo et al. (41)),(▲) Charleston, WV (42), (△) Portland, OR (21), and (■) Boston, MA (3, 43). See Table IV, Footnote a for Refs. to coal data.

the various elements after release from the coal-fired plants. In fact, some studies cited above show that enriched elements from coal-fired plants are preferentially associated with fine particles relative to Al and other lithophile elements. If the larger particles bearing the lithophile elements preferentially settle out of the atmosphere, the EF values of the enriched elements will increase during transit to receptor sites. We examine this possibility below.

One of our aims has been to observe systematic behaviors of elements released on particles from sources of a given class that would allow one to predict compositions of particles released from various members of that class. As more coal-fired plants have been studied than other types of sources, they provide a good opportunity to search for systematic behavior. Because of great variations of ash content of coals, types of boilers and pollution control devices, it would be very difficult to predict absolute emission rates. However, one can often bring order into many kinds of data by normalization to the crustal abundance pattern, i.e., by using enrichment factors. Column 2 of Table IV shows average EF values and their standard deviations for all available data on coal-fired plants equipped with electrostatic precipitators (ESPs). We have not included the limited data on plants with scrubbers, as recirculation of the scrubber water plus additions of acid-neutralizing chemicals confuse interpretations of the emissions (44). Also, we have confined the elements in Table IV mainly to those that are enriched in urban atmospheres, ignoring many lithophile elements with EF values near unity.

The results shown in Col. 2 of Table IV are rather discouraging in that, for most elements, the σ value is comparable to the mean value. Much of the variation may be due to variations of the compositions of the input coals. All but one plant operate in the United States, but some use eastern U.S. coals and other consume western U.S. coals, which usually have large differences in the concentrations of many elements (17, 45). One could perhaps remove this source of fluctuation by calculating the EF values with respect to the input coals rather than to the crust. Thus, in Col. 3, we list the EF$_{coal}$ values, defined as

$$EF_{coal} = (C_x/C_{Al})_{particles}/(C_x/C_{Al})_{coal} \qquad (3)$$

where the denominator refers to the coal consumed at the plant studied.

The use of EF$_{coal}$ values reduces the relative fluctuations of several elements: V, Mn, Zn, As, Se, Mo, Cd, Sb and Pb. However, in most cases, the reduction is not very dramatic and, for other elements (Cr, Co, Ni, Cu), the fluctuations become worse.

Another source of fluctuation is the efficiency curve of the ESP. An ESP that lets through large amounts of large, aluminosilicate particles will cause reductions of the EF values relative to one that discriminates more strongly against large particles.

Table IV.   Enrichment Factors of Elements on Particles from Coal
            Combustion[a]

Element	All Particles				Fine Particles Only			
	EF_{crust}		EF_{coal}		EF_{crust}		EF_{coal}	
V	3.6±2.5	(8)[b]	2.4±1.2	(9)[b]	2.7±0.9	(4)[b]	3.3±2.3	(4)[b]
Cr	6.1±6.6	(8)	3.4±4.1	(9)	1.2±1.0	(4)	1.8±1.2	(4)
Mn	0.53±0.36	(7)	1.25±0.53	(8)	0.52±0.34	(4)	1.7±0.5	(4)
Co	2.7±0.8	(7)	2.0±0.9	(8)	1.9±0.8	(3)	2.9±1.9	(4)
Ni	9±6	(4)	10±15	(4)	2.0±1.7	(3)	2.3±0.7	(3)
Cu	8±4	(6)	2.1±1.8	(5)	9±6	(3)	2.5±0.1	(2)
Zn	18±24	(9)	5±2	(10)	7±1	(4)	7±4	(4)
As	160±150	(8)	5.3±3.9	(10)	270±430	(4)	16±8	(4)
Se	920±1100	(9)	5.4±5.1	(9)	680±530	(4)	11±11	(4)
Mo	200±180	(5)	7.2±3.5	(6)	60±40	(3)	8±6	(3)
Cd	130±160	(3)	10±8	(4)	25	(1)	5	(1)
Sb	85±40	(8)	5±2	(10)	130±65	(3)	13±9	(3)
Pb	15±12	(7)	9±7	(6)	10±6	(3)	7±2.5	(3)

[a]Based on data from Refs. 5, 6, 7, 18, 27-36.   Includes data only
 for plants with electrostatic precipitators, not scrubbers.
[b]Numbers in parantheses indicated number of plants included.

Thus, fluctuations might be reduced by considering only particles
of a specified size range.   In Cols. 4 and 5, we show the EF_{crust}
and EF_{coal} values calculated only for fine particles, here taken
as particles of diameters less than about 2.5 µm.   Surprisingly,
the EF values for fine particles are, in many instances, smaller
than the corresponding values for particles of all sizes in Cols.
2 and 3, despite the fact that the enrichment of many of the elements listed on fine particles is well established and rather well
explained by vaporization/ recondensation mechanisms (18, 34).
This result may be deceptive, however, as we had to eliminate the
many studies from Cols. 2 and 3 that contained no size information, so the two sets of averages are not for the same populations
of plants.   A further problem with these data may also be the inclusion of particles up to 2.5¬µm diam, whereas, recent studies
(44, 46) indicate that the large changes of composition occur below about 0.5¬µm diam. Thus, our "fine" particles in Table IV include particulate mass between 0.5 and 2.5 µm that is chemically
similar to larger particles, which reduces the EF values. We made
the cut at 2.5 µm to develop components for separate CEBs of fine
and coarse particles in ambient air, which are commonly divided at
about that point.
    Unfortunately, confining the EF values to fine particles does
not cause a dramatic reduction of the fluctuations, and they become worse for some elements.   Fluctuation may still be present
because of differences in types of coal, boilers, ESPs and operating conditions. If data were available on a large number of coal-
fired plants, one could perhaps sort them into a dozen or so categories,   each   having rather   narrow   ranges   of   these plant

parameters. Then, perhaps, the ranges of $\underline{EF}_{coal}$ or $\underline{EF}_{crust}$ values for elements from plants in each category would have little fluctuation. At present, data are not available for nearly enough plants to attempt this kind of treatment.

Another approach might be to use even more fundamental treatments of particle formation in combustion sources. We have used rather simple treatments based on enrichment factors. Other groups (e.g., Ref. 18) have attempted more complex models based on vaporization of elements or their compounds in the combustion zone, recondensation of many species on particle surfaces as the temperature of the gas stream is reduced at later stages in a plant, and coagulation of the very fine particles. It is not clear if that approach could be used to make practical, reliable predictions of compositions of particles of various sizes released from a plant. If so, an accurate knowledge of many parameters would be needed, e.g., the composition of the coal (including the distribution of some elements in different phases), particle size of the coal as burned, residence times and temperatures throughout the plant, and the efficiency of the ESP versus particle size. Ideally, the calculation should be extended to cover additional condensation and coagulation that occur outside of the stack in the early part of the plume.

Comparisons with Ambient Fine Particles. Above we noted a potential problem of using $\underline{EF}_{crust}$ values to determine possible coal contributions to ambient levels of various elements, namely that large, aluminosilicate particles might preferentially settle out of the atmosphere between the source and receptor. Here we have tried to avoid that problem by calculating $\underline{EF}$ values only for fine particles from several cities and from coal-fired power plants as shown in Table V. Again, fine particles are those with diam < 2.5 µm. The $\underline{EF}$s for coal in Table V are taken from Col. 4 of Table IV.

Are the $\underline{EF}$ values for coal great enough for any of these elements that coal combustion can be the major source of the element in these cities? This condition is met only for Fe and, possibly, for V in Charleston. Aside from these possiblities and As and Se, it is unlikely that major amounts of these elements are contributed by coal combustion. Arsenic and Se are special cases. In the CEBs discussed above, As is an important marker for coal and much of the predicted Se arose from that source in the Washington area. Within the large limits of error on the $\underline{EF}_{crust}$ value for As on fine particles from coal combustion, one can account for the values for ambient particles in Washington and Denver; however, the value for Charleston is far greater than for coal-fired plants. Possibly the observed value for Charleston is wrong, as the As K x-ray line has strong interference from a Pb L x-ray in XRF determinations. Lewis and Macias (42) considered the As data so uncertain that they did not include As in their factor analysis. A second problem is that As is moderately volatile - not as

Table V.   Enrichment Factors for Fine Particles with Respect to Earth's Crust[a]

Element	Charleston[b]	Portland[c]	Washington[d]	St. Louis[e]	Denver[f]	Coal[g]
V	<9	45	140	–	24	2.7
Cr	–	54	80	–	18	1.2
Mn	4.3	24	5.5	8	7	0.5
Fe	1.8	2.4	1.9	1.6	1.1	~1
Ni	–	180	–	21	30	2
Cu	290	240	210	360	115	9
Zn	230	350	710	550	235	7
As	6,400	–	450	–	420	270
Se	35,000	–	12,000	8,200	60,000	680
Cd	–	–	–	8,200	–	130
Sb	–	–	5,800	–	–	130
Pb	19,000	13,000	–	10,000	23,000	10

[a]Renormalized so that avg. of five lithophiles = 1.00.
[b]Ref. (42).   [c]Ref. (21).   [d]Ref. (16).
[e]Based on our analysis of 1976 data from Station 105, see Ref. (41).
[f]The 1978 Denver Winter Haze (S. L. Heisler et al., Env. Research and Technology, Inc., 1979).
[g]From last column of Table IV.

as much so as the halogens and Se, but a portion of it appears to be in the gas phase at stack temperatures (47). Also, much of the in-stack data were obtained with cascade impactors. Because of bounce-off problems, some of the large-particle lithophiles were probably deposited on the small-particle stages and the back-up filter, thus reducing the EF values. By contrast, most data on ambient particles (the exception being Washington) were obtained with virtual impactors, which suffer little bounce-off.

The enormous enrichments of Se on ambient particles relative to particles from coal-fired plants are surely due to the same problems as noted for As, but especially the well established fact that a major fraction of Se is in the vapor phase at stack temperatures (47, 48). The very large Se enrichments for ambient particles indicate that much of the vapor-phase Se condenses on particles after release. In some areas there may be additional important sources of As and Se, e.g., non-ferrous smelters.

In summary, coal-fired power plants appear not to be the major source of most enriched elements on particles in urban areas, despite the great attention devoted to mechanisms by which those elements become preferentially attached to fine particles (e.g., Refs. 18, 34). However, the detailed studies of processes in coal-fired plants are of considerable value, as the fundamentals should be applicable to other kinds of combustion sources. Furthermore, it may be necessary to use this fundamental approach to develop methods for predicting the source compositions for coal-fired power plants that have not been measured. Selenium is much

more enriched on ambient particles than on particles from coal-
fired plants, probably because the Se is mostly in the gas phase
at stack temperatures, but condenses on particles in ambient air.
To a much lesser extent, As may have a similar behavior.

Implications for Future Source Measurements

As discussed above, many detailed studies of sources are
needed to establish compositions of source components for
receptor-model use. Coal-fired power plants have been studied
more than other sources, but even in that case, there are not
nearly enough data to establish trends for predicting components
for plants that have not been studied. The situation is much
worse for most other sources, except possibly refuse incinerators,
of which the three studied release particles of very similar compositions (9, 10).

When sources are studied, several things should be done to
provide data needed for receptor-model applications. First, particles should be collected in at least two different size fractions corresponding to the division at about 2.5-μm diam now used
in many studies of ambient aerosols. In some cases, it may be desirable to have more size cuts. As noted above, compositions of
particles from coal combustion change dramatically below about
0.5-μm diam (44, 46). Above we identified a minimum of about
twenty elements that should be measured. Also, in order to develop adequate markers for sources that emit carbonaceous particles, measurements of organic compounds and other properties related to carbonaceous particles should be made.

Better collection methods are needed for stack measurements.
Virtual impactors or other devices that avoid the bounce-off problems of cascade impactors should be developed. Methods for collection of very large amounts of size-segregated particles suitable for detailed organic analyses are needed. Better low-blank
filter and collection-surface materials that can withstand high
stack temperatures are needed. Careful attention should be given
to the determination of vapor-phase species of volatile elements
and compounds.

Although in-stack measurements will continue to be important,
they are not totally adequate for determining components appropriate to receptor models. Ideally, one would rather know the sizes
and compositions of particles far enough from the stack that any
condensible species would have condensed, very large particles
would have fallen out, and coagulation of fine particles would
have occurred. These measurements should be coordinated with in-
plant studies for comparison with the calculations suggested
above. More instrumental developments are needed for sampling in
plumes, as one would like to use high volume samplers that collect
size-segregated samples isokinetically. To our knowledge, not all
of those conditions have been met in aircraft sampling (e.g.,
Refs. 7, 12). Although one normally thinks of airplanes for plume

studies, it is very difficult to meet those conditions above the
stall speeds of airplanes. Perhaps it would be worthwhile to develop special platforms for plume sampling, e.g., lighter-
than-air, manned balloons or tethered balloons with especially designed samplers.

   Not all emissions come from stacks. Often the unducted fugitive process emissions and entrained dust are very important, but
there are no good ways of studying them. Perhaps the best way
would be the use of ground-level samplers controlled by wind-
sector devices that turn on the samplers only when the wind is in
the desired direction and above some minimum velocity. The wind-
trajectory method noted below may also be of value for study of
fugitive emissions.

   One particular type of source that should be studied carefully is entrained soil. As shown above, this is often the greatest contributor of TSP in urban areas. As there is so much of it
present, we need to know concentrations of all measured elements
quite well to make an accurate determination of the residual
amounts left to be accounted for by other sources. The composition of seived soil is often used for the soil component, but
there may be considerable fractionation imposed by entrainment,
e.g., preferential selection of very fine clay mineral particles.
Such fractionation has been demonstrated in the very limited studies of entrainment of particles from soil of known composition
(e.g., Refs. 21, 49). These studies can probably best be done in
controlled environments such as wind tunnels. One cannot simply
collect ambient particles in the countryside and consider it to be
soil, as there are anthropogenic contributions even at great distances from cities (15). There is further confusion betwen clean,
"continental" dust and "urban" dust. The latter, which is usually
collected near city streets (21, 50), typically has a composition
of soil contaminated by anthropogenic emissions, especially from
motor vehicles.

   Finally, we should note a new method that may be of considerable use for obtaining components for certain sources from net-
works of ambient samplers: the wind-trajectory method (51). We
are developing this approach with the use of data on about twenty
elements measured in the fine and coarse fractions from a network
of ten dichotomous samplers in and near St. Louis for two years in
connection with the Regional Air Pollution Study (RAPS) (41),
which also included collection of good wind data. Data at each
station are analyzed to determine overall average concentrations
and standard deviations for each element in each size fraction.
Then we search the data for sampling periods during which wind
fluctuations were small ($\sigma_\theta < 20$ deg) and concentrations of the elements were very high, typically about 3σ above average. For each
element we plot a histogram of the mean wind directions for all
samples meeting these criteria. Often these histograms have clus-
ters about certain wind directions that point towards dominant
point sources of the element.

This procedure is carried out for each element at each site. Then we construct maps of the trajectories from the various stations for each element. Often, the trajectories from several stations intersect at the location of the dominant source for that element, which we can identify from source maps for the St. Louis area. Once all samples that are heavily influenced by emissions from that source are identified, we determine relative concentrations of other elements associated with it by performing linear regressions of all elements with the dominant element.

This method for obtaining components for receptor models has several advantages relative to other methods. First, the component will be determined at the receptor sites, after any condensation, coagulation or fallout during transit has occurred. Second, it may include fugitive emissions from the source as well as ducted emissions. Third, the measurements are made at the same time and with the same device as the ambient sampling. Fourth, the measurements are made on emissions from specific sources in the local area, not just on the same class of source somewhere else. Fifth, the measurements do not require cooperation of the source operators or intrusion upon their property.

The wind trajectory method is a supplement to other methods, not a replacement, as it has some weaknesses. Although we are already certain of being able to pick up emissions from Cu and Zn smelters, a pigment plant and steel plants, the method may not work on sources that have no dominant element. It may not work on widely distributed, ubiquitous sources such as gravel quarries, or on sources with very tall stacks. Despite these limitations, the method will be a useful supplement to the others and it should be fully exploited.

In this paper, we have focussed on the weaknesses of our present knowledge about the compositions of particles from sources that are needed as input for receptor models. However, despite these weaknesses, we feel that the receptor model is probably already capable of more accurate determinations of TSP contributions from various types of sources than the classical methods of source emissions inventories coupled with dispersion models. If the measurements suggested are made, then the receptor models should provide very accurate estimates of those contributions.

Acknowledgments

This work was in part supported by the NSF/RANN Program under Grant No. ENV75-02667 and by the U.S. Environmental Protection Agency under Cooperative Agreement CR 806263-02-0. Computer time was in part supplied by the Maryland Computer Center.

Literature Cited

1. Gordon, G. E. Environ. Sci. Technol. 1980, 14, 792.
2. Friedlander, S. K. Environ. Sci. Technol. 1973, 7, 235.

3. Hopke, P. K.; Gladney, E. S.; Gordon, G. E.; Zoller, W. H.;
    Jones, A. G. Atmos. Environ. 1976, 10, 1015.
4. Alpert, D. J.; Hopke, P. K. Atmos. Environ. 1980, 14, 1137.
5. Gladney, E. S.; Small, J. A.; Gordon, G. E.; Zoller, W. H.
    Atmos. Environ. 1976, 10, 1071.
6. Gladney, E. S. Ph.D. Thesis, Univ. of Maryland, 1974.
7. Small, J. A. Ph.D. Thesis, Univ. of Maryland, 1976.
8. Mroz, E. J. Ph.D. Thesis, Univ. of Maryland, 1976.
9. Greenberg, R. R.; Zoller, W. H.; Gordon, G. E. Environ. Sci.
    Technol. 1978, 12, 566.
10. Greenberg, R. R.; Gordon, G. E.; Zoller, W. H.; Jacko, R. B.;
    Neuendorf, D. W.; Yost, K. J. Environ. Sci. Technol. 1978,
    12, 1329.
11. Germani, M. S.; Small, M.; Zoller, W. H.; Moyers, J. L.
    "Fractionation of Elements During Copper Smelting", Environ.
    Sci. Technol. (in press).
12. Small, M.; Germani, M. S.; Small, A.; Zoller, W. H.; Moyers,
    J. L. "An Airborne Plume Study of Emissions from the
    Processing of Copper Ores in Southeastern Arizona", Environ.
    Sci. Technol. (in press).
13. Small, M. Ph.D. Thesis, Univ. of Maryland, 1979.
14. Ondov, J. M.; Zoller, W. H.; Gordon, G. E. "Composition of
    Particles Emitted by Motor Vehicles" (in preparation).
15. Thomae, S. C. Ph.D. Thesis, Univ. of Maryland, 1977.
16. Kowalczyk, G. S. Ph.D. Thesis, Univ. of Maryland, 1979.
17. Germani, M. S.; Gokmen, I.; Sigleo, A.; Kowalczyk, G. S.;
    Olmez, I; Small, A.; Anderson, D. L.; Failey, M. P.;
    Gulovali, M. C.; Choquette, C. E.; Lepel, E. A.; Gordon, G.
    E.; Zoller, W. H. Anal. Chem. 1980, 52, 240.
18. Smith, R. D. Prog. Energy Combust. Sci. 1980, 6, 53.
19. Kowalczyk, G. S.; Choquette, C. E.; Gordon, G. E.;
    Atmos. Environ. 1978, 12, 1143.
20. Kowalczyk, G. S.; Rheingrover, S. W.; Gordon, G. E.
    "Chemical Element Balances of Atmospheric Particles in the
    Washington, D.C. Area", submitted to Environ. Sci. Technol.
21. Watson, J. G., Jr. Ph.D. Thesis, Oregon Graduate Center,
    1979.
22. Cooper, J. A.; Watson, J. G., Jr. "Review of Air Particulate
    Source Impact Assessment Methods", This Symposium.
23. Winchester, J. W. N.Y. Acad. Sci. 1980, 338, 190.
24. Daisey, J. M.; Kneip, T. J. "Atmospheric Particulate Organic
    Matter: Multivariate Models for Identifying Sources and
    Estimating Their Contributions to the Ambient Aerosol", This
    Symposium.
25. Hites, R. A. "Sources and Fates of Atmospheric Polycyclic
    Aromatic Hydrocarbons" This Symposium.
26. Novakov, T. "Role of Carbon Soot in Sulfate Formation",
    presented at the ACS/CSJ Chemical Congress, Honolulu,
    Apr. 1979.

Kaakinen,

27. Kaakinen, J. W.; Jorden, R. M.; Lawansani, M. H.; West, R. E. **Environ. Sci. Technol.** 1975, **9**, 862.
28. Klein, D. H.; Andren, A. W.; Carter, J. A.; Emergy, J. F.; Feldman, C.; Fulkerson, W.; Lyon, W. G.; Ogle, J. C.; Talmi, Y.; VanHook, R. I.; Bolton, N. W. **Environ. Sci. Technol.** 1975, **9**, 973.
29. Block, C.; Dams, R. **Environ. Sci. Technol.** 1976, **10**, 1011.
30. Mesich, F. G.; Schwitzgebel, K. "Coal Fired Power Plant Trace Element Study - A Three Station Comparison", Radian Corp., Austin, TX, Report No. TS-1a, 1975.
31. Gorman, P. G.; Shannon, L. J.; Schrag, M. P.; Fiscus, D. E. "St. Louis Demonstration Final Report: Power Plant Equipment, Facilities and Environmental Evaluations", U.S. EPA Report EPA-600/2-77-155b, 1977.
32. Ondov, J. M.; Ragaini, R. C.; Biermann, A. H. **Environ. Sci. Technol.** 1979, **13**, 598.
33. Ragaini, R. C.; Ondov, J. M. "Trace Contaminants from Coal-Fired Power Plants", in **Proc. of the Int. Conf. on Environmental Sensing and Assessment, Las Vegas, Sept. 1975** (IEEE, New York, 1976) Paper 17-2.
34. Natusch, D. F. S.; Wallace, J. R.; Evans, C. A., Jr. **Science** 1974, **183**, 202.
35. Coles, D. G.; Ragaini, R. C.; Ondov, J. M.; Fisher, G. L.; Silberman, D.; Prentice, B. A. **Environ. Sci. Technol.** 1979, **13**, 455.
36. Smith, R. D.; Campbell, J. A.; Nielson, K. K. **Environ. Sci. Technol.** 1979, **13**, 553.
37. Gordon, G. E.; Zoller, W. H.; Gladney, E. S. "Abnormally Enriched Trace Elements in the Atmosphere", in "Trace Substances in Environmental Health - VII", D. D. Hemphill, ed.; Univ. of Missouri, Columbia, MO, 1973, pp. 167-174.
38. Rahn, K. A. "The Chemical Composition of the Atmospheric Aerosol", Technical Report, Grad. School of Oceanography, Univ. of Rhode Island, July, 1976.
39. Wedepohl, K. H. In "Origin and Distribution of the Elements", L. H. Ahrens, ed.; Pergamon Press, London, 1968, pp. 999-1016.
40. Moyers, J. L.; Ranweiler, L. E.; Hopf, S. B.; Korte, N. E. **Environ. Sci. Technol.** 1977, **11**, 789.
41. Loo, B. W.; French, W. R.; Gatti, R. C.; Goulding, F. S.; Jaklevic, J. M.; Llacer, J.; Thompson, A. C. **Atmos. Environ.** 1978, **12**, 759.
42. Lewis, C. W.; Macias, E. S. **Atmos. Environ.** 1980, **14**, 185.
43. Wiltsee, K. W., Jr.; Cogley, D. R. "Analysis of Total Suspended Particulate Concentration in Metropolitan Boston", Walden Division of Abcor, Inc., June, 1977.
44. Ondov, J. M.; Biermann, A. H. "Elemental Composition of Atmospheric Fine Particle Emissions from Coal-Combustion in a Modern Utility Electric Power Plant", This Symposium.

45.   Gluskoter, H. J.; Ruch, R. R.; Miller, W. G.; Cahill, R. A.;
      Dreher, G. B.; Kuhn, J. K. "Trace Elements in Coal:
      Occurrance and Distribution", Ill. State Geol. Survey
      Circular 499, 1977.
46.   Flagen, R. C.; Taylor, D. D. "Aerosols from a Laboratory
      Pulverized Coal Combustor", This Symposium.
47.   Germani, M. S. Ph.D. Thesis, Univ. of Maryland, 1980.
48.   Andren, A. W.; Klein, D. H.; Talmi, Y. Environ. Sci.
      Technol. 1975, 9, 856.
49.   Miller, M. S.; Friedlander, S. K.; Hidy, G. M. J. Colloid
      Interface Sci. 1972, 39, 165.
50.   Hopke, P. K.; Lamb, R. E.; Natusch, D. F. S. Environ. Sci.
      Technol. 1980, 14, 164.
51.   Rheingrover, S. W.; Gordon, G. E. " Identifying Locations of
      Dominant Point Sources of Elements in Urban Atmsopheres from
      Large, Multi-Element Data Sets", in Proc. of the 4th Int.
      Conf. on Nuclear Methods in Environmental and Energy
      Research, Univ. of Missouri, Columbia, MO, Apr., 1980, J. R.
      Vogt, ed. (in press).

RECEIVED March 25, 1981.

Review of the Chemical Receptor Model of Aerosol Source Apportionment

JOHN A. COOPER

Department of Environmental Science, Orgeon Graduate Center, 19600 N.W. Walker Road, Beaverton, OR 97006

There are two general types of aerosol source apportionment methods: dispersion models and receptor models. Receptor models are divided into microscopic methods and chemical methods. Chemical mass balance, principal component factor analysis, target transformation factor analysis, etc. are all based on the same mathematical model and simply represent different approaches to solution of the fundamental receptor model equation. All require conservation of mass, as well as source composition information for qualitative analysis and a mass balance for a quantitative analysis. Each interpretive approach to the receptor model yields unique information useful in establishing the credibility of a study's final results. Source apportionment sutdies using the receptor model should include interpretation of the chemical data set by both multivariate methods.

Urban aerosols are complicated systems composed of material from many different sources. Achieving cost-effective air particle reductions in airsheds not meeting national ambient air quality standards requires identification of major aerosol sources and quantitative determination of their contribution to particle concentrations. Quantitative source impact assesment, however, requires either calculation of a source's impact from fundamental meteorological principles using source oriented dispersion models, or resolving source contributions with receptor models based on the measurement of characteristic chemical and physical aerosol features. (1)

Although source oriented dispersion models are invaluable predictive tools, their ability to quantify the impact of a source is
limited. Receptor oriented methods of source apportionment, however, have evolved in recent years to the point where they now
clearly form a new discipline of air pollution science. (1,2) This
new discipline is distinctly different from dispersion modeling and
has demonstrated that it can quantitatively apportion source contributions to particle levels.   (3-7) Receptor models are not predictive tools and as such, have minimal applicability in estimating
the effectiveness of future control strategies. These two models,
however, are complementary in nature. Receptor models, for example,
can provide a precise quantitative determination of the contribution each source type made to air particulate levels, the results
of which can be used to calibrate a dispersion model to provide the
highest level of confidence in its predictions. (8)

A variety of receptor oriented source apportionment tools have
been developed, which can be grouped into two general categories as
illustrated in Figure 1. (1) Microscopic methods include both optical and electron microscopic approaches which use morphology,
color, and elemental content to qualitatively identify particles,
while particle volume, density and number must be estimated for a
quantitative analysis. The microscopic approach is the older of
the two general receptor methods and an extensive library of "microscopic fingerprints" consisting of morphological, color and elemental features has been developed over the past few decades. Microscopic methods are limited primarily by their relatively poor
precision and the high cost associated with analyzing a sufficient
number of particles to adequately represent the entire group of
particles collected on a filter.

The chemical method of receptor modeling is the most recent approach to source apportionment (1,4,9,10) and extensive libraries
of source "chemical fingerprints" have yet to be established. Many
different chemical methods have evolved over the past 10 to 15 years.
Included in this category are chemical mass balance (CMB), factor
analysis (FA), multiple regression, cluster analysis, enrichment
factors, pattern recognition, principal component analysis, time
series and spatial distribution analysis, target transformation
factor analysis, etc. These methods of interpreting chemical data
have evolved from different origins and are often perceived as distinctly different models when, in most cases, their only major differences are terminology and approach.

Chemical methods, in general, identify aerosol sources by comparing ambient chemical patterns or fingerprints (interelemental
patterns, spatial or time variant patterns) with source chemical
patterns. Source contributions are quantified by a least squares
multiple regression analysis on either the total mass on different
filters or the mass of individual chemical species on a single filter. Although similarities in the different chemical approaches
are greater than their differences, they have been historically
divided into two categories: chemical mass balance methods, which

attempt to define the most probable linear combination of sources
to explain the chemical pattern on a single filter; (1,4,11) and
multivariate methods, which attempt to define the most probable
linear combination of sources to explain either the time or spatial
variability in ambient chemical patterns. (4,12-14)

Each source apportionment tool (Figure 2) has its unique
strengths and limitations, and each can provide valuable insight
into sources contributing to air particulate levels. The most cost
effective tool or set of tools, however, will depend on the nature
of the airshed, potential sources and the accuracy and precision of
source apportionment required.

The information provided by these models is circumstantial in
nature and the results from a single interpretive approach at this
stage of model evolution may be insufficient to develop the level
of confidence required to support strong action or clear decisions.
The objective of source apportionment studies must be to build a
strong enough bridge of circumstantial information (Figure 3) to
quantitatively relate a source to an impact.   Thus, the entire information base must support and be internally consistent with a
study's conclusions to provide decision makers with confidence that
their actions will result in improved air quality.

Recent publications (1,4,9,10) have provided extensive reviews
of receptor models. The objective of this paper is to discuss selective aspects of the chemical receptor model.

The Chemical Receptor Model

Source-dispersion and receptor-oriented models have a common
physical basis. Both assume that mass arriving at a receptor (sampling site) from source j was transported with conservation of mass
by atmospheric dispersion of source emitted material. From the
source-dispersion model point of view, the mass collected at the
receptor from source j, M_j, is the dependent variable which is equal
to the product of a dispersion factor, D_j (which depends on wind
speed, wind direction, stability, etc.) and an emission rate factor,
E_j, i.e.,

$$M_j = D_j E_j$$

From the receptor model viewpoint, the total aerosol mass, M,
collected on a filter at a receptor is the dependent variable and
equal to a linear sum of the mass contributed by p individual sources,

$$M = \sum_{j=1}^{p} M_j$$

The mass of individual chemical species, m_i, is also assumed to be
a linear sum of the contributions of element i from each source,

$$m_i = \sum_{j=1}^{p} F_{ij} M_j \tag{1}$$

where F_{ij} is the fraction of the ith chemical species in emissions
from the jth source. Equation (1) can be transformed to a fractional mass concentration form by dividing both sides of equation (1)

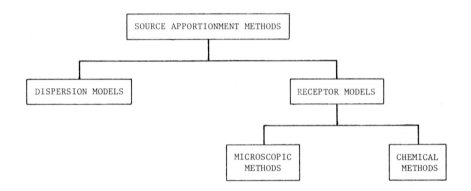

Figure 1.   Schematic comparison of source apportionment methods

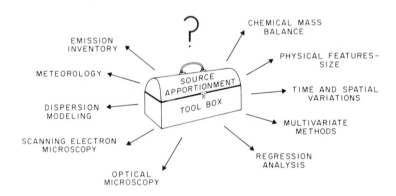

Figure 2.   Which tools should be used?

by the total deposit mass, M, multiplying by 100% and generalizing
for the kth filter as shown in the following equation,

$$C_{ik} = \sum_{j=1}^{p} F_{ij} S_{jk} \qquad (2)$$

where C_{ik} is the percent concentration of the ith chemical species
on the kth filter and S_{jk} is the percent of the total mass on the
kth filter contributed by the jth source. The kth filter may be
either one in a series of filters collected during different time
intervals at one site or one in a series collected during the same
time interval at different sites. The structure of the matrices
noted in equation (2) are illustrated in Figures 4 and 5.

It needs to be emphasized at this point that a model is a mathematical representation of the real world. If two models have the
same mathematical representation of the real world, they are, in
fact, the same model. Chemical mass balance, principal component
factor analysis, target transformation factor analysis, etc. have,
for all practical purposes, identical mathematical representations
(Equation 1) of the real world and start with the same input data
matrices (Figure 4). The principal difference in these "different"
receptor models is their approach to the solution of either Equation (1) or Equation (2).

The chemical mass balance method starts with a single column
vector from the ambient data matrix, C_k. This vector represents
the chemical concentrations for the kth filter, which is combined
with the best available estimates of the source compositions from
the fractional composition matrix, F_{ij}, to form a series of linear
equations in which the M_j are the only unknowns. This set of equations is then solved by the least squares method to obtain the best
fit of the ambient chemical data on a single filter.

Multivariate methods, on the other hand, resolve the major
sources by analyzing the entire ambient data matrix. Factor analysis, for example, examines elemental and sample correlations in the
ambient data matrix. This analysis yields the minimum number of
factors required to reproduce the ambient data matrix, their relative chemical composition and their contribution to the mass variability. A major limitation in common and principal component factor analysis is the abstract nature of the factors and the difficulty these methods have in relating these factors to real world
sources. Hopke, et al. (13,14) have improved the methods' ability
to associate these abstract factors with controllable sources by
combining source data from the F matrix, with Malinowski's target
transformation factor analysis program. (15) Hopke, et al. (13,14)
as well as Kleinman, et al. (10) have used the results of factor
analysis along with multiple regression to quantify the source contributions. Their approach is similar to the chemical mass balance
approach except they use a least squares fit of the total mass on
different filters instead of a least squares fit of the chemicals
on an individual filter.

Each method of data interpretation provides its own unique

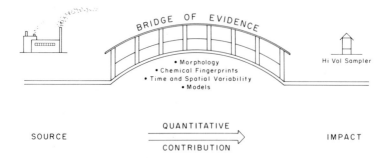

Figure 3. Circumstantial source apportionment information. Contribution based on establishing other circumstances that afford reasonable inference of the source contribution.

1. Ambient Chemical Data Set

$$C_{n\times m} = \begin{vmatrix} C_{11} & C_{12}\cdots C_{1k}\cdots C_{1m} \\ C_{21} & \\ \vdots & \\ C_{i1} & C_{i2}\cdots C_{ik} \\ \vdots & \\ C_{n1} & C_{n2}\cdots C_{nk}\cdots C_{nm} \end{vmatrix} = \left[C_{ik}\right]_{n\times m} = \begin{vmatrix} C_{Na,1} & C_{Na,2}\cdots C_{Na,m} \\ C_{Mg,1} & \\ C_{Al,1} & \\ \vdots & \\ C_{Pb,1} & \end{vmatrix}$$

(Time or Location →; Chemical Species ↓)

2. Source Composition Data Set

$$F_{n\times p} = \begin{vmatrix} F_{11} & F_{12}\cdots F_{1j}\cdots F_{1p} \\ F_{21} & \\ \vdots & \\ F_{i1} & F_{i2}\cdots F_{ij} \\ \vdots & \\ F_{n1} & F_{n2}\cdots F_{nj}\cdots F_{np} \end{vmatrix} = \left[F_{ij}\right]_{n\times p} = \begin{vmatrix} F_{Na,Auto} & F_{Na,Marine} & F_{Na,Kraft}\cdots \\ F_{Mg,Auto} & & \\ F_{Al,Auto} & & \\ \vdots & & \\ F_{Pb,Auto} & & \end{vmatrix}$$

(Source →; Chemical Species ↓)

Figure 4. Input data

insight into the nature of the aerosol sources. Which method will
be most effective in resolving specific sources will depend on the
airshed, sources of interest, and their characteristics. The method
of data interpretation, however, should not necessarily be limited
to a single approach but should include both mass balance and multivariate approaches which should yield the highest level of confidence in a study's conclusion.

Conservation of Mass

Source-dispersion and receptor-oriented models both assume that
mass is conserved in transport of material from source to receptor.
The validity of this assumption is a matter of degree and its utility depends on the specific source, airshed and model. The problems
associated with this assumption are illustrated in Figure 6. Material emitted from a source can be either in the gaseous or aerosol
phase. It may go through a number of chemical and physical changes
before it is collected on a filter and measured. A portion of the
aerosol phase, for example, may evaporate before it is collected or
particles may be removed through sedimentation in transport. The
gaseous phase, on the other hand, may contribute to the aerosol deposited on the filter by condensation or through atmospheric chemical reactions. In addition, filter artifact effects and evaporative
losses may alter the material deposited on the filter before it is
weighed.
  These potential problems appear to substantially limit the validity of the conservation of mass assumption. Its validity, however,
is a matter of degree and will depend on the specific source and how
well the potential events noted above have been minimized through
experimental design. The effect of these potential chemical and
physical changes may be reduced by sampling source emissions with
size selective samplers to minimize the effects due to changes in
chemistry from sedimentation, and by using dilution sampling to mit-
igate the effect of condensation and evaporation. ($\underline{16}$-$\underline{18}$) This,
plus selection of appropriate filters to limit artifact effects and
a thorough knowledge of potential chemical reactions can minimize
the effects of deviations from conservation of mass.
  A substantial amount of confusion ($\underline{9}$,$\underline{10}$,$\underline{13}$,$\underline{14}$) has recently developed as to an approach's dependence on conservation of mass. As
Cooper and Watson ($\underline{1}$) have noted, the F_{ij} factors refer to the source
chemistry as it arrives at the receptor. It is assumed with the conservation of mass that the F_{ij} as might be measured at a receptor,
is the same as have been measured at the source. As noted above,
this may not be valid depending on the source and the method used
for source sampling. The chemical mass balance method incorporates
the F_{ij} directly in its calculations and as a result is often perceived as having a greater dependence on this assumption than methods
such as factor analysis which do not use F_{ij} values in their calculations. Factor analysis methods, however, identify abstract factors,
which explain variability. It is impossible to attribute a common

$$
S_{p \times m} = \begin{vmatrix} S_{11} & S_{12} & \cdots S_{1k} & \cdots S_{1m} \\ S_{21} & & & \\ \vdots & & & \\ S_{j1} & S_{j2} & \cdots S_{jk} & \cdots S_{jm} \\ \vdots & & & \\ S_{pl} & S_{p2} & \cdots S_{pk} & \cdots S_{pm} \end{vmatrix} = \begin{bmatrix} S_{jk} \end{bmatrix}_{p \times m} = \begin{vmatrix} \overset{\text{Time or Location} \longrightarrow}{S_{Auto,1}} & S_{Auto,2} & \cdots S_{Auto,m} \\ S_{Marine,1} & & \\ S_{Kraft,1} & & \\ \vdots & & \\ S_{p,1} & & \end{vmatrix} \Bigg\downarrow \text{Source}
$$

Figure 5.   Unknown

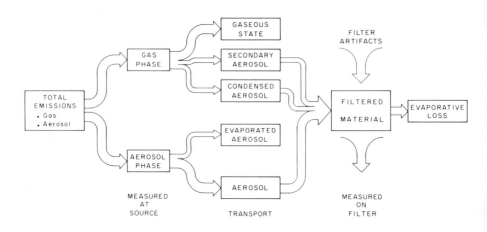

Figure 6.   Aerosol mass balance

source name to these abstract factors without additional information. The F_{ij} values derived from factor analysis, for example,
have little relevance unless they can be associated with known or
familiar F_{ij} values derived from prior studies of source emissions.
    Examples of the dependence of factor analysis on prior know-
ledge of the chemistry of potential sources is clearly illustrated
in the literature. Blifford and Meeker, (19) who were the first to
apply factor analysis to an aerosol data set, resolved seven major
factors from a set of five-year average elemental compositions from
thirty cities. The largest factor found contained Fe, Mn and Ti
and explained 21% of the variability. They attributed this source
to heavy industry based on their prior knowledge of source chemistry
in 1967. Although they considered the possibility of assigning this
source to soil because of the strong Ti dependence, they ruled it
out because this factor was highest in cities such as Birmingham,
Cleveland and Pittsburgh, was not correlated with cities expected
to have high wind blown dust, and was independent of population
class. This factor, however, would probably be attributed today
to road dust with the current knowledge of source and ambient aerosol chemistry and the typical level of impact from road dust. In
addition, Hopke, et al. (20) in their factor analysis of a Boston
data set, found a factor rich in Mn and Se for which they could not
assign a common source identity because they had no knowledge of a
source rich in these two elements.
    More recent publications using factor analysis have recognized
the above limitations, (13,14) but have concluded that less source
information is required by factor analysis than chemical mass balance methods. It has also been suggested that factor analysis is
less dependent on the conservation of mass assumption because it
examines the variability in the chemistry at the source and is
therefore less dependent on what happens between source and receptor. This concept, however, fails to recognize that the usual objective of a source apportionment study is to develop sufficient
"confidence" in the results to stimulate "effective action" to reduce ambient particulate levels. Assume, for example, that the relative chemistry of material emitted by a specific source is substantially altered in transit to a receptor and that the effects of
this source have been resolved by factor analysis. How much "confidence" can be attributed to an assignment of the resolved factor to
this particular source, if the factor's chemistry differs substantially from what is known about the source's chemistry? The degree
of confidence in source assignment is directly related to the degree
of similarity between the factor's and the source's chemical composition.
    The level of confidence in source assignment can be increased
if other observables such as time variability, size dependence, etc.
are consistent with prior knowledge of a source's emission characteristics. Confidence can also be enhanced by showing that the
characteristics of other possible sources are inconsistent with the
ambient observables. It follows that the highest level of confidence

84   ATMOSPHERIC AEROSOL

in a study's conclusions will result when the characteristics of
all possible sources in an airshed are known. These latter points
emphasize the circumstantial nature of receptor model results which
may require additional information to support the "level of confidence" required to attain the most effective action.

The Study and Source Resolution

Urban aerosols are complex mixtures of chemicals contributed
by many different sources. One objective of an aerosol source apportionment study is to separate or resolve the contribution of specific sources or source types from the collection of all possible
sources, and quantify their contributions.
There are three main phases of a source apportionment study,
which includes sampling, analysis and interpretation. Each must be
optimized to attain the highest level of source resolution and most
accurate apportionment. The first phase, sampling, is as critical
as the other two in contributing to the overall resolving power of
a study. The sampling parameters which exert the greatest influence
on source resolution, include selection of sampler, filter, sampling
frequency and sampling duration. These parameters, however, are not
independent and their choice will represent a compromise between
maximizing analytical sensitivity, precisely defining the system's
variability and available resources.
A dichotomous sampler with a fine to coarse cut point of about
2 μm is preferred because of the bimodal nature of the ambient aerosol and its sources. This type of physical separation according to
particle size, is the first stage of source resolution since it separates the particles as well as sources into two broad classes.
This particle sorting removes chemical interferences and increases
analytical sensitivity. It also increases the level of confidence
when most of a road dust contribution is, for example, found in the
coarse fraction.
The filter material of choice is a thin teflon membrane since
it minimizes artifact formation and maximizes analytical sensitivity
by X-ray fluorescence analysis. Although X-ray fluorescence (XRF)
may not be the only analytical technique used, it is generally accepted as being the most cost effective analysis for source apportionment. (2) Its background and therefore, analytical sensitivity,
is dependent on the filters' surface density. The analytical sensitivity of XRF for aerosols deposited on a stretched teflon membrane with a density of about 0.3 to 0.4 mg/cm^2, for example, is
about three times greater than an aerosol deposited on a cellulose
based filter with a surface density of about 4 mg/cm^2. This difference can be translated into either more information for the same
analytical costs or the same information for a lower analysis cost.
Selection of sampling frequency and duration will be determined
primarily by the relevant standard, potential sources and resources.
The current particulate standard is based on a 24-hour and annual
average. Thus, sampling duration should probably not exceed 24 hours.

Sampling for less than 24 hours should substantially improve source resolution due to more precisely defined time dependence and enhanced signal to noise ratios. Sampling for less than 24 hours does increase cost substantially, and it may not be necessary for the first phase of a study.

The primary objective of the chemical analysis step is to accurately measure the major chemical components and key indicating species such as Pb for automotive exhaust, Ni and V for residual oil, Al and Si for road dust, etc. Although it is not essential that all of the major chemical species be measured, it greatly improves the credibility and confidence in the final results if most of the aerosol mass is explained. Carbon and silicon should be measured because they represent two of the most abundant chemical species present in a typical urban aerosol, (21) and because they are useful indicating elements (oxygen, which may at times be the most abundant element, is impractical to measure). In general, Na, Mg, Al, S, K, Ca, Fe and Pb should be measured because of their abundance and their roll in source fitting. Other elements such as P, Cl, Ti, V, Cr, Mn, Ni, Cu, Zn, Br, Rb, Sr, Zn, Cd, In, Sn, Ba, rare earths, etc. explain smaller portions of the total mass but may be key indicating elements. Chemical species such as ammonium and nitrate ions, and specific organic compounds, may also be useful. Measurement of specific organic compounds should be undertaken only after a thorough review of the expected source, transport, and analysis chemistries has shown the candidate compounds to have a reasonably constant production rate, a reasonable aerosol lifetime and measured by a relatively low cost analysis technique. The set of essential chemicals and preferred analytical techniques will be determined by the specific sources in each airshed. Selection of an optimum sampling and analysis protocol, based on a thorough understanding of the chemistry of potential sources, can result in substantial cost savings and greatly improved source resolution.

The interpretation stage consists of applying one or all of the chemical receptor model approaches to interpreting the chemical data generated. The objective of a source apportionment study is the support of effective control action. The level of confidence required to initiate this action may be established with a single receptor model interpretive approach or it may require information from additional interpretive approaches, wind sector analysis, (4, 22) dispersion models, microscopy, etc.

The first step of any source apportionment study should therefore include a thorough review of potential sources, their chemistry and time dependence, if the highest level of confidence is to be established in the final results.

Conclusions

The chemical receptor model is one of the most precise tools currently available for assessing the impact of aerosol sources. Each interpretive approach to the receptor model yields unique information useful in establishing the credibility of the final results.

A source apportionment study using the receptor model should include interpretation of the chemical data set by both multivariate and chemical mass balance methods. The most critical steps in a receptor model study are the initial review of potential source characteristics and the development of an appropriate study plan.

Acknowledgments

Many helpful discussions with Richard T. DeCesar are gratefully acknowledged. This work was partially funded by a grant from the U.S. Environmental Protection Agency (R807375010).

References

1. Cooper, J.A.; Watson, J.G., Jr. "Receptor Oriented Methods of Air Particulate Source Apportionment." J. Air Pol. Cont. Assoc., 1980, 30, 1116-1125.
2. Watson, J.G. "The State of the Art of Receptor Models Relating Ambient Suspended Particulate Matter to Sources." Draft Final Report to Environmental Protection Agency, Environmental Research and Technology, Concord, Mass. 01742.
3. Cooper, J.A.; Watson, J.G., Jr.; Huntzicker, J.J. "Summary of the Portland Aerosol Characterization Study." Presented at the 1979 Annual Air Pollution Association Meeting in Cincinnati, Ohio.
4. Watson, J.G., Jr. "Chemical Element Balance Receptor Model Methodology for Assessing the Sources of Fine and Total Suspended Particulate Matter in Portland, Oregon." Ph.D Dissertation, Oregon Graduate Center, Beaverton, OR 1979.
5. Cooper, J.A.; Watson, J.G., Jr. "Portland Aerosol Characterization Study." Final Report to the Oregon State Department of Environmental Quality, July, 1979.
6. DeCesar, R.T.; Cooper, J.A. "Medford Aerosol Characterization Study." Final Report to the Oregon State Department of Environmental Quality, February, 1981.
7. NEA Laboratories, Inc. Potlatch Corporation Aerosol Characterization Study, Final Report for Test Period Jan., 1979 - Dec., 1979. April, 1980.
8. Core, J.E.; Hanrahan, P.L.; Cooper, J.A. "Air Particulate Control Strategy Development: A New Approach Using Chemical Mass Balance Methods." This symposium.
9. Gordon, G.E. "Receptor Models." Environ. Sci. Technol., 1980 14, 792-800.
10. Kleinman, M.T.; Pasternack, B.S.; Eisenbud, M.; Kneip, T.J. "Identifying and Estimating the Relative Importance of Sources of Airborne Particles." Environ. Sci. Technol., 1980, 14, 62.
11. Friedlander, S.K. "Chemical Element Balances and Identification of Air Pollution Sources." Environ. Sci. Technol., 1973, 7, 235.
12. Henry, R.C. "A Factor Model of Urban Aerosol Pollution." Ph.D. Dissertation, Oregon Graduate Center, Beaverton, OR 1977.

13. Hopke, P.K. "Source Identification and Resolution Through Application of Factor and Cluster Analysis." Annals of the New York Academy of Sciences, ed. K. Kneip, 1980.
14. Alpert, J.D.; Hopke, P.K. "A Quantitative Determination of Sources in the Boston Urban Aerosol." Atmos. Environ., 1980, 14, 1137-1146.
15. Malinowski, E.R.; Howery, D.G.; Weiner, P.H.; Soroka, J.M.; Funke, P.T.; Selzer, R.B.; Levinstone, A.R. "FACTANL-Target Transformation Factor Analysis," Quantum Chemistry Program Exchange, Indiana University, Bloomington, Inc., 1976, Program 320.
16. Heisohn, R.J.; Davis, J.W.; Knapp, K.T. "Dilution Source Sampling Systems," Environ. Sci Technol., 1980, 10, 1205.
17. Williamson, A.D.; Smith, W.B. "A Dilution Sampling System for Condensation Aerosol Measurements: Design Specifications," Southern Research Institute, Special Report #SORI-EAS-79-793, 1979.
18. Houck, J.E.; Larson, E.R.; Cooper, J.A. "A Dilution Source Sampler for Developing Chemical Mass Balance Fingerprints." (In preparation).
19. Blifford, I.H.; Meeker, G.O. "A Factor Analysis Model of Large Scale Pollution." Atmos. Environ. 1967, 1, 147.
20. Hopke, P.K.; Gladney, E.S.; Gordon, G.E.; Zoller, W.H.; Jones, A.G. "The Use of Multivariate Analysis to Identify Sources of Selected Elements in the Boston Urban Aerosol." Atmos. Environ. 1976, 10, 1015-1025.
21. Cooper, J.A. "Urban Aerosol Trace Element Ranges and Typical Values." Battelle Pacific Northwest Lab. Tech. Publications BNWL-SA-4690, 1973.
22. Gordon, G.E.; Zoller, W.H.; Kowalczyk, G.S.; Rheingrover, S.W. "Composition of Source Components Needed for Aerosol Receptor Models." (This symposium).

RECEIVED April 21, 1981.

The State of the Art of Receptor Models Relating Ambient Suspended Particulate Matter to Sources

JOHN G. WATSON and RONALD C. HENRY— Environmental Research and Technology, Inc., Concord, MA 01742

JOHN A. COOPER—Oregon Graudate Center, 19600 N.W. Walker Road, Beaverton, OR 97006

EDWARD S. MACIAS—Washington University, St. Louis, MO 63130

Receptor models are used to determine the source contributions to ambient particulate matter loadings at a sampling site based on common properties between source and receptor. This is in contrast to a source model which starts with emission rates and meteorological measurements to predict an ambient concentration.

Three generic types of receptor model have been identified, chemical mass balance, multi-variate, and microscopical identification. Each one has certain requirements for input data to provide a specified output. An approach which combines receptor and source models, source/receptor model hybridization, has also been proposed, but it needs further study.

The input to receptor models is obtained from ambient sampling, source sampling, and sample analysis. The design of the experiment is important to obtain the most information for least cost. Sampling schedule, sample duration and particle sizing are part of the ambient sampling design. Analysis for elements, ions, carbon, organic and inorganic compounds are included in the sample analysis design. Which sources to sample and how to sample them are part of the source sampling design.

In order for receptor modeling to become a useful tool, it must be developed in four major areas:

1. General Theory - The generic types of receptor model are related to each other, but that relationship has not been generally established. A general theory of receptor models, including the input data uncertainties, needs to be constructed.

0097-6156/81/0167-0089$05.00/0

2. Validation - Simulated data sets created
    from known source contributors and per-
    turbed by random error should be presented
    to models, and their source contribution
    predictions should be compared to the known
    contributions. Several models should be
    applied to the same data set and their
    results compared.

3. Standardization- Validated models need to be
    placed in standard form for easy implementation
    and use.

4. Documentation and Education - Users need to
    be informed of the powers and limitations of
    these models and instructed in their use.

There is a need today to quantify the effects of aerosol
sources on ambient particulate matter loadings. Identifying the
major sources of ambient particulate matter loadings was a fairly
simple process when values exceeded 500 $\mu g/m^3$ and stack
emissions were plainly visible. Control of these emitters was
forthcoming and effective. At levels of 150 to 200 $\mu g/m^3$, the
use of annual emission inventories focused further regulatory
efforts on major sources which have resulted in more successful
reductions. Presently, at levels around 75-100 $\mu g/m^3$, the
uncertainties involved in these assessments of source
contributions are greater than the contributions themselves.

Source-oriented atmospheric dispersion modeling has been the
major tool used in attributing ambient concentrations to source
emissions. With the development of inexpensive and rapid
chemical analysis techniques for dividing ambient and source
particulate matter into its components has come another approach,
the receptor model.

While the source-oriented model begins with measurements at
the source (i.e., emission rates for the period under study), and
estimates ambient concentrations, the receptor-oriented model
begins with the actual ambient measurements and estimates the
source contributions to them. The receptor model relies on
properties of the aerosol which are common to source and receptor
and that are unique to specific source types. These properties
are composition, size and variability.

The examination of these properties to identify source
contributions is not new. Throgmorton and Axetell[1] have done
a good   job of tabulating and generalizing applications of
receptor models in the past and Cooper and Watson[2] and
Gordon[3] offer excellent reviews of current work. Receptor
modeling is an active area of research which is becoming a
useful tool in the ambient monitoring and regulatory process.
However, it needs a unified focus to direct its development.

In an attempt to provide this focus, forty-seven active receptor model users from government, university, consulting and industry met for 2 1/2 days in February 1980[4]. They addressed the models and the information required to use them in six separate task forces: 1) Chemical Element Balance Receptor Models, 2) Multivariate Receptor Models, 3) Microscopic Identification Receptor Models, 4) Field Study Design and Data Management, 5) Source Characterization, and 6) Analytical Methods. The objectives of these interrelated task forces were to:

1. define the state-of-the-art via summaries of past and present research,
2. outline future research required to make the models useful tools for the scientific community, regulators, and planners,
3. suggest vehicles for carrying out that research within the context of ongoing projects.

This paper presents a summary of the results and recommendations of this meeting. A complete report is available elsewhere.

Types of Receptor Models

Receptor models presently in use can be classified into one of four categories: chemical mass balance, multivariate, microscopic, and source/receptor hybrids. Each classification will be treated individually, though it will become apparent that they are closely related.

The starting point for the receptor model is the source model. Though the source model may not deliver accurate results under many conditions, its limitations are primarily due to its inability to include every environmentally relevant variable and inadequate measurements for the variables it does include. The general mathematical formulations, however, are representative of the way in which particulate matter travels from source to receptor.

Each pollution source's contribution to a receptor sample is the product of an emissions factor and a dispersion factor in the source model formulation. The total concentrations measured at the receptor is the linear sum of this product.

At the receptor, the mass concentration of aerosol property i, C_i, will be

$$C_i = \sum_{j=1}^{p} a_{ij} S_j \qquad (1)$$

where a_{ij} is the mass fraction of source contribution S_j possessing property i <u>at the</u> receptor.

<u>The Chemical Mass Balance Receptor Model.</u> Equation 1 which

can be derived from a source model, is Friedlander's (5) chemical element balance. Another name for this model might be
chemical mass balance. Historically, elemental measurements were
most readily available. Recent analytical developments allow
carbon compounds and ions to be among the properties considered.

If one measures n chemical properties of both source and
receptor, n equations of the form of Equation 1 exist. If the
number of source types contributing those properties is less than
or equal to the number of equations, i.e., p < n, then the
source contributions, the S_j, can be calculated.

Four methods of performing this calculation have been
proposed, the tracer property, linear programming, ordinary
linear least squares fitting, and effective variance least
squares fitting.

The tracer property method is the simplest. It assumes that
each aerosol source type possesses a unique property which is
common to no other source type. It works well when a) the
concentration of the tracer property in the source material is
well known and invariant; b) the concentration of the tracer
property can be measured accurately and precisely in the ambient
sample; c) the concentration of the tracer property at the
receptor comes only from one source type.

These conditions cannot be completely met in practice, and
by limiting the model to only one tracer property per source
type, valuable information contained in the other aerosol
properties is being discarded. Thus, solutions to the set of
equations like Equation 1 have been developed to make use of
the additional information provided by more than one unique
chemical property of a source type, and even that of properties
which are not so unique.

The linear programming and weighted least squares solutions
deal with the case of a number of constraints, n, greater than
the number of unknowns, p. Henry (6) has applied a linear
programming algorithm (7). Results are comparable to those
obtained from ordinary weighted least squares. This solution
has not been developed further.

In the ordinary weighted least squares method, the most
probable values of source contributions are achieved by
minimizing the weighted sum of squares of the difference between
the measured values of the ambient concentration and those
calculated from Equation 1 weighted by the analytical uncertainty
of those ambient measurements. This solution provides the added
benefit of being able to propagate the measured uncertainty of
the ambient concentrations through the calculations to come up
with a confidence interval around the calculated source
contributions.

Dunker (8) and Watson (9) have applied the treatment of
Britt and Luecke (10) to the solution of a set of Equations 1.
This method weights the difference between measured and

calculated concentrations by the uncertainties of ambient and source measurements.

This solution provides two benefits. First, it propagates a confidence interval around the calculated source contributions which reflects the cumulative uncertainty of the input observables. The second benefit provided by this "effective variance" weighting is to give those chemical properties with larger uncertainties, or chemical properties which are not as unique to a source type, less weight in the fitting procedure than those properties having more precise measurements or a truly unique source character.

Each least squares solution exhibits instabilities when source compositions are similar, though not exact. Very large or negative source contributions result.

The input data required for the chemical mass balance consist of the following:

1. The concentrations of the chemical components which comprise the total mass of the sample, and their precisions.

2. The number and identification of source types contributing to the receptor samples.

3. The source type compositions and their variabilities.

Obtaining this input data is not a trivial matter. The greatest progress in recent years has been made in obtaining the data in requirement 1, the ambient chemical component concentrations.

Meeting the second data requirement is more difficult. In fact, one of the objectives of the receptor model must be to identify as well as to quantify source contributions. Some, such as geological material, automobile exhaust, oil and wood combustion may be located throughout the airshed. Watson (9) included these first in his Portland calculations. Gaps between the calculated and measured aerosol property concentrations suggest other sources, which he included in a new calculation, which yielded new gaps, suggesting new sources, etc. Although the results are reasonable, they are also subjective.

Ambient aerosol property measurements are quite sophisticated; source measurements are not as well developed. Certain aerosol property measurements have been made on emissions of many sources. Unfortunately these tests seldom provide a full accounting of samples, sometimes do not report confidence intervals on the measurements, and often do not fully describe the operating parameters of the source at the time of the test.

The source contributions of aerosol formed from gaseous emissions, such as sulfate, nitrate and certain organic species, cannot be quantified by chemical mass balance methods. Watson (9) proposes a unique source type which will put an upper limit on the contributions of secondary aerosol sources, but it cannot attribute those contributions to specific emitters.

The future development of the chemical mass balance receptor
model should include 1) more chemical components measured in
different size ranges at both source and receptor; 2) study of
other mathematical methods of solving the chemical mass balance
equations; 3) validated and documented computer routines for
calculations and error estimates; and 4) extension of the
chemical mass balance to an "aerosol properties balance" to
apportion other aerosol indices such as light extinction.

Multivariate Models. Multivariate methods can incorporate
the variability of ambient concentrations and source emissions
which mass balance methods can not do at the present time.
Linear regression (11,12,13), correlation (14,15) and factor
analysis (6,16) are the forms these models take.
        While the chemical mass balance receptor model is easily
derivable from the source model and the elements of its solution
system are fairly easy to present, this is not the case for
multivariate receptor models. Watson (9) has carried through
the calculations of the source-receptor model relationship for
the correlation and principal components models in forty-three
equation-laden pages.
        The mathematical obfuscation of these models must not remove
the requirement that every receptor model must be representative
of and derivable from physical reality as represented by the
source model. A statistical relationship between the variability
of one observable and another is insufficient to define cause
and effect unless this physical significance can be established.
        Multivariate models have been successful in identifying
source contributions in urban areas. They are not independent
of information on source composition since the chemical component
associations they reveal must be verified by source emissions
data. Linear regressions can produce the typical ratio of
chemical components in a source but only under fairly restrictive
conditions. Factor and principal components analysis require
source composition vectors, though it is possible to refine these
source composition estimates from the results of the analysis
(6,17).
        The most important caveat to be heeded in the use of multi-
variate models is that though origination in the same source
will cause two chemical components to correlate, the converse,
that chemical components which correlate must have originated
in the same source, is not true.
        The input data required of the multivariate receptor models
at their present state-of-the-art are the ambient concentrations.
The output with this input, however, is only qualitative. With
an estimate of the a_{ij}, the source compositions and contributions
can be refined. No provision has yet been made for propagating
the measurement uncertainties through the model or of evaluating
their effect on model output as has been done for the chemical
mass balance.

Future development of the multivariate models should include testing of other multivariate methods. Suggestions include ridge regression (18), regression against principal components (19), clustering techniques (20), time series analysis (26), and non-parametric (22) approaches. Applied mathematicians and aerosol specialists should work together on these models to become aware of the full range of multivariate methods available and to apply them appropriately to air quality problems.

Further work ought to include creation of simulated ambient data sets from a simple source model with known source compositions and contributions perturbed by typical experimental error. The multivariate models should be applied to these data sets, as did Watson (9) for the chemical mass balance, to determine the extent to which the models can apportion source contributions under various conditions.

Microscopic Identification Models. Many different optical and chemical properties of single aerosol particles can be measured by microscopic identification and classification in order to distinguish particles originating in one source type from those originating in another. The microscopic analysis receptor model takes the form of the chemical mass balance equations presented in Equation 1.

$$M_i = \sum_{j=1}^{p} a_{ij} \, S_j \tag{2}$$

where

M_i = % mass of particle i in the receptor sample
a_{ij} = fractional mass of particle type i in source type j
S_j = % of total mass measured at the receptor due to source j.

The microscopic receptor model can include many more aerosol properties than have been used in the chemical mass balance and multivariate models. The data inputs required for this model are the ambient properties measurements and the source properties measurements. To estimate the confidence interval of the calculated source contributions the uncertainties of the source and receptor measurements are also required. Microscopists generally agree that a list of likely source contributors, their location with respect to the receptor, and windflow during sampling are helpful in confirming their source assignments.

The major limitation of microscopic receptor models is that the analytical method, the classification of particles possessing a defined set of properties, has not been separated from the source apportionment of those particles. Equation 2 has never been used in this application. The source identification takes place by recognition of the particle by the microscopist and thus is difficult to standardize.

Source contributions assigned to the same aerosol sample have varied greatly in intercomparison studies (23); but, without the intermediate particle property classifications, it is impossible to ascribe the differences to the analytical portion or to the source assignment portion of the process. Future development for this model includes creation and acceptance of a standardized methodology, definition of important optical criteria and their measurement methods, and a method of reporting and cataloging source properties by means of the chosen criteria.

Hybrid Source/Receptor Models. Until now, the receptor models have been treated as if they were completely separate entities from the source models. This need not be the case. A source model incorporates measured or estimated values for an emission rate factor and the dispersion factor. Whenever either of these enter the receptor model as observables, we call it a hybrid model. The three applications considered here are emission inventory scaling, micro-inventories, and dispersion modeling of specific sources within a source type.

Emission inventory scaling, proposed by (24), uses the relative emission rates of two source types subject to approximately the same dispersion factor (e.g., residential heating by woodstoves and natural gas) to approximate the source contribution from the source type not included in the chemical mass balance (e.g., natural gas combustion). The ratio of the emission rates is multiplied by the contribution of the source type which was included in the balance.

The micro-inventory approach developed by Pace (25) has been shown to be a good predictor of annual average TSP based on a detailed assessment of emissions close to the sampling site. Four types of emissions related to distance from a sampling site are compiled using standard emissions factors where applicable. Pace investigated many other variables in multiple linear regressions with TSP but found these to be the significant ones.

The micro-inventory should be a prerequisite for aerosol mass balance or microscopic model applications to identify the most likely sources for sampling, analysis and inclusion in the balance.

A major limitation of receptor models is their inability to distinguish between specific sources within a source type. Resuspended road dust may be a major cause of standard violations, but until the offending roadways can be pinpointed, a control strategy cannot be implemented. A major limitation of source models is the necessity to estimate emission rates from the many, diverse producers of suspended particulate matter. The receptor model quantifies the source type contributions. Only the major contributors need to be evaluated for the source model, so that resources which might have been used to inventory an entire airshed can be concentrated on determining emission rates from the specific emitters within a source type. This improved data with fewer source inputs should increase the accuracy with which the relative contribution from each individual source type can be calculated by the source model. An application of this hybridization has never been attempted, but it merits exploration.

Two more sets of observables are introduced into the hybrid models ; the emissions factors and the dispersion factors. It is the difficulty of quantifying these that led to the use of a receptor model over the source model in the first place, so it would seem there is little advantage in reintroducing them. The advantage of the hybridization is that the number of individual emission and dispersion factors can be considerably reduced and that the relative values rather than the absolute values are used. These relative values are more accurate in most cases. Still, the uncertainties of emission and dispersion factors need to be evaluated and incorporated into any source/receptor hybrid model.

Future areas of development include theoretical formulation of source/receptor models and validation tests with real and simulated data.

Receptor Model Input Data

The measurements required for the present receptor models include particulate matter composition, size and variability for both source and receptor. Obtaining these data requires attention to field study design and data management, source characterization, and analytical methods.

Field Study Design and Data Management. Few aerosol characterization studies performed in the past have been designed with the goal of applying a receptor model. The receptor models have traditionally been developed and applied after data collection as a method of interpreting large, complicated sets measurements. This has been a productive development mechanism, but the present state-of-the-art demands a more precise definition of the goals of a study, the receptor models to be used, the input data required, and the measurement and quality assurance program to obtain that input data.

Cost will be the primary controlling factor in the experimental design in essentially all cases. The cost of obtaining the entire population of possible measurements is prohibitive. The field study design must select a subset which is representative of the population within definable confidence limits and which can be obtained within the constraints of existing resources.

Several generalizations can be made about aerosol characterization study design based on past experience. Existing
data should be used to obtain an understanding of the area under
study. These data include historical aerosol and gas concentrations, meteorological data, emissions inventories, chemical and
microscopical analyses, and the results of dispersion modeling.

Siting of monitoring stations should take advantage of
existing sites with consideration to land use categories and
transportation source influences. Source-oriented sites are
often useful, but both upwind and downwind, baseline or background sites are also needed. Source-oriented modeling can
provide helpful guidance for the location of sites.

The frequency and duration of measurement are important
factors; source contributions at a fixed station are often
short-term and to measure their variability requires sampling
times of the same order as that variation. Twenty-four hour
sampling is considered minimal, and shorter term sampling is
often desirable, particularly when periods of constant wind
direction are required. Measurements are needed to extrapolate
to annual average conditions; however, it is not necessary to
sample daily to achieve this. It is important to obtain
information on both intense pollution conditions (episodes) and
relatively clean conditions. Thus, more samples than needed
should be taken with a subset chosen for detailed analysis based
on meteorological or other factors. Samples should be taken
every day over selected seasonal periods rather than one per
third or sixth day to capture the progression effect of multiday
events.

Meteorological information which has been useful to receptor
model studies includes temperature, relative humidity, mixing
height, windspeed and wind direction. The wind direction is
particularly important for the verification of certain receptor
model source contribution predictions; definite differences
should exist between samples on which the source is upwind and
downwind of the receptor.

Source characterization must be an integral part of any
field study design. A microinventory (25) of each sampling
site should be conducted and specimens of nearby fugitive
emissions sources should be taken and analyzed. Aerosol properties of likely point source contributors identified in the
emission inventory must be measured during the course of
ambient sampling. Source sampling and analysis techniques need
to be compatible with ambient techniques.

An overwhelming number of chemical compounds can be found
in the typical urban aerosol, and some subset must be selected
for quantification. Two general rules can narrow the scope of
chemical analysis somewhat. First, the sum of the masses of
chemical components measured should equal, within stated
precision estimates, the total mass concentration of the aerosol.
Without this equivalence one cannot necessarily assign the

unaccounted for mass to a source, nor is it possible to verify the other component quantities via a mass balance. Thus the first requirement of chemical analysis is that the major chemical components of total aerosol mass in the sample are quantified.

The second requirement is that components unique to specific aerosol sources affecting the ambient sample are measured. The precise chemical species will depend on the sources in the airshed under consideration. Most studies will require measurement of certain "basic" particulate matter species including ions, elements, and organic and inorganic carbon.

A quality assurance component must be integral to the study to verify the accuracy of its measurements and to estimate a precision for each one. This component should include co-located sampling, replicate analysis, station audits, data validation, interlaboratory comparisons and a full set of standard operating procedures. The quality of the data set generated should be discussed and should include the results of validation, analysis of outliers and overall estimates of accuracy and precision. Every data item should be considered an interval rather than a number, and the precision of each measurement determined from calculations (27).

The final important ingredient of experimental design is data management, which is normally carried out by computer. The data must be organized and merged into uniform formats accessible to the receptor models. Provisions for the entry and processing of quality assurance data will greatly reduce the efforts required for this often neglected study component. All data from initial analyses must be retained until the sorting of relevant and irrelevant information can be achieved. Economical accessibility, security and definition of uncertainty should be keynotes of data management systems for this work. Organization techniques for recovering subsets of data from a number of large data bases to a microcomputer must be developed to maximize economic data handling and computation.

Future development efforts in field study design and data management should 1) develop methodologies for choosing sampler location, sampling schedule, and sampling devices; 2) measure the collection efficiency of available samplers as a function of particle size, windspeed and wind direction; 3) create a series of "typical" design scenarios which can be specified for different aerosol study objectives; 4) develop standardized data bases from which ambient and source information can be recovered. 5) develop standardized receptor models for computer implementation and standard interfaces to the relevant data bases; and 6) create quality assurance procedures and methods of reducing the data to assign a confidence interval to each measured quantity.

Source Characterization. All receptor models, even the source/receptor hybrids, require input data about the particulate matter sources. The multivariate models, which can conceivably be used to better estimate source compositions, require an initial knowledge of the chemical species associations in sources.

Existing data on characteristics of particles from various types of sources are inadequate for general use, though they have been used in specific studies with some success. Most of the source tests have been made for purposes other than receptor modeling and complete chemical and microscopical analyses have not been performed. Source operating parameters which might affect the aerosol properties of emissions have not been identified nor measured in ambient sampling and no provision is made for likely transformations of the source material when it comes into equilibrium under ambient conditions.

For receptor modeling, the only source characteristics that are relevant are those perceived at the receptor. But if source components could be separated from each other at the receptor, there would be no need for a receptor model.

Stack sampling is not sufficient for characterizing all sources and may not be the best method even for ducted point sources. Because of condensation of vapors, fallout of very large particles and the neglect of non-ducted emissions, the materials collected from stacks, especially those at high temperatures, will often not represent the particles from the source observed at a receptor several kilometers away.

For use with receptor models, it is desirable to collect particles far enough from the source to allow the emissions to reach equilibrium with the atmosphere, but not so far away that the source material under test becomes contaminated with aerosol from other sources.

Armstrong et al. (28) describe a unique approach of attaching sampling equipment to tethered balloons which are maneuvered from the ground into the plume of smokestack emissions. Aircraft sampling in plumes (29) has long been used, but it is expensive and difficult.

A promising development is that reported by Rheingrover and Gordon (30). Using ambient samplers located in "maximum effect" areas for specific sources and wind flow patterns, they have distinguished the composition of a single point source at a receptor. They then use these source characteristics to represent that source in areas less affected by that source, where its contribution is not so obvious.

In-stack testing procedures are still under development with a greater inclination on the part of developers to produce samples of use for receptor modeling. New sampling methods include dilution with clean air to simulate ambient conditions, filter media amenable to chemical analyses, and samples in size ranges similar to those sampled in ambient air.

Source characterization results are not located in a centralized facility which is constantly updated. The Environmental Protection Agency has established the Environmental Assessment Data System (EADS) (31) which contains chemical compositions of particulate matter emissions tests. This existing computerized structure can provide the centralized location for receptor model source characterization information. Procedures such as those described for ambient data in the previous section need to be developed in order to allow receptor model users access to this data base over telephone lines. The data required of receptor model source tests should be incorporated into the EADS, and source characterization results should report this information in an EADS compatible format.

Future development efforts in source characterization should 1) develop new source sampling methods including tethered balloon and ground based sampling; 2) standardize data reporting and management procedures and store validated data in a central data base; and 3) create a chemical component analysis protocol to obtain maximum information from each source test.

Analytical Methods. Many chemical and physical analysis methods exist to characterize particulate matter collected on a substrate. Though several methods are multi-species, able to quantify a number of chemical components simultaneously, no single method is sufficient to both quantify the majority of the collected particulate matter mass and those components which serve to identify and quantify source contributions.

The range of analytical methods available for particulate matter analysis is large; many of the available methods are critically reviewed by (32). A complete review of all methods applied in the past with a view toward their use for receptor modeling does not exist, and its production would be a useful contribution to the state-of-the-art. Several analytical tools have been found to meet the stated criteria, or are in the development stages of trying to meet them, and are being used to supply the input measurements for receptor models. One means of classifying available and useful analysis procedures for source and receptor studies consists of the following categories: 1) elements with atomic number greater than eleven; 2) carbon, 3) ions; 4) organic compounds; 5) inorganic compounds; and 6) physical and optical properties.

Development areas for analytical methods applied to receptor models should include the following:
1. Critically review all aerosol analysis methods for their practicality to receptor modeling. Determine the extent to which they meet the necessary and desirable criteria. Identify potentially useful methods for development.

2. Develop standard reference materials representing
    different ambient and source particulate matter
    matrices. Characterize them in terms of relevant
    aerosol properties.
3. Standardize and conduct intermethod comparison tests
    of organic compound, inorganic compound, carbon, and
    microscopical analysis methods.

Furthering the State-of-the-Art

Receptor modeling is a new and growing field. Unlike many
fields of science which start with a few key researchers and
branch out to include new ones, receptor modeling's roots are
spread over an array of disciplines and organizations. Nuclear
spectroscopists, chemical engineers, industrial plant managers,
regulatory agency personnel — each has approached the
relationship of source to receptor by aerosol properties from his
own point of view. It is not surprising that a coherent theory
of receptor modeling and standardized methodologies have yet to
emerge.

Each of the study areas summarized in the previous sections
has ended with a list of research recommendations. These lists
are not exhaustive, but they do contain those issues which the
most experienced scientists in the field of receptor modeling
believe are in need of resolution for the specific study area.
For receptor modeling in general, four major development areas
exist.

The primary need of receptor modeling today is a general
theory within which to operate. Each specific receptor model
application should be derivable from this framework. If not,
either the application of the theory is incorrect, or the theory
must be changed. Throgmorton and Axetell (1) have compiled the
applications. Henry(6) and Watson (9) have outlined the
theory including some of those applications. The theory must be
expanded to include them all.

Next, the applications have to be validated and placed into
standardized forms. Validation should consist of two steps.
First, simulated data sets of aerosol properties should be
generated from pre-selected source contributions as did
Watson(9) in his simulation studies of the chemical mass
balance method. These data should be perturbed with the types of
uncertainties expected under field conditions. The types of
sources and their contributions predicted by the receptor model
application should be compared with the known source model values
and the extent of perturbation tolerable should be assessed.
This will define the precision required of the input data for the
application in question.

The next step in validation requires a real data set with as
many properties of source and receptor as can be obtained in an

area with well-known sources. All receptor models should be applied and the results should be examined for a self-consistent picture of source contributions. If such a picture does not emerge, the inconsistent applications must be reformulated or discarded. A systematic elimination of measurements from the data set should then be made to find the most cost-effective sampling and analysis scheme to provide valid information within a specified confidence interval.

The third need is standardization. The receptor model applications need to be written as standard computer routines, common data structures that can accommodate uncertainties of the observables need to be created, and sampling and analysis equipment and procedures must produce equivalent results. Field study "scenarios" for typical situations should be proposed.

The fourth and final need is for documentation and education. The validation and standardization will go for naught if the practice of receptor modeling cannot be established at the state implementation plan level where it is most sorely needed. Major reviews of model applications, analytical methods, source characterization and field study design need to be prepared and communicated to those most likely to make use of them.

The mechanisms for carrying out the research required are not clear. These mechanisms must be cost-effective while involving as wide an array of talent as possible. It is evident from the meeting of experts which generated this report that cross-disciplinary and inter-institutional approaches to meeting these needs must be fomented.

Enough receptor modeling by other names is going on today that, with the proper coordination and leadership, the cost of pursuing these developments should not be prohibitive. Most of the development should take place within the context of ongoing projects.

Acknowledgements

The workshop on which this report was based was jointly sponsored by the Industrial Environmental Research and Environmental Science Research Laboratories of EPA under the direction of John Milliken and Robert Stevens. Participants and major contributors other than the authors to the ideas expressed in this article are Bill Baosel, John Bachman, Neil Berg, Glenn Cass, John Core, Russ Crutcher, Lloyd Currie, Joan Daisey, Stuart Dattner, Briant Davis, Ron Draftz, Alan Dunker, Tom Dzubay, Bob Eldred, Ed Fasiska, Sheldon Friedlander, Don Gatz, Glen Gordon, Bruce Harris, George Hidy, Phil Hopke, Reginald Jordon, Ted Kneip, Nick Kolak, Ross Leadbetter, Richard Lee, Chuck Lewis, Carol Lyons, John Milliken,

Jarvis Moyers, John Overton, Tom Pace, Skip Palenki,
Terry Peterson, Bill Pierson, Ken Rahn, Phil Russell, John
Spengler, Robert K. Stevens, Warren White, Jack Winchester,
John Woodward and John Yocom. Appreciation is due to Peter
K. Mueller, Steven L. Heisler and Judith C. Chow for their
helpful comments on the presentation of this subject.

Literature Cited

1. Throgmorton, J.A. and Axetell, K. (1978). Digest of
    Ambient Particulate Analysis and Assessment Methods, EPA-
    450/3-78-113, Research Triangle Park, North Carolina.

2. Cooper, J.A. and Watson, J.G. (1980). "Receptor Oriented
    Methods of Air Particulate Source Apportionment," Journal
    of the Air Pollution Control Association, 30, 1116.

3. Gordon, G.E. (1980). "Receptor Models," Environmental
    Science & Technology, Inc., 14, 795.

4. Watson, J.G. et al. (1980). "The State of the Art of
    Receptor Models Relating Ambient Suspended Particulate
    Matter to Sources", EPA-Research Triangle Park, North
    Carolina.

5. Friedlander, S.K. (1973). "Chemical Element Balances and
    Identification of Air Pollution Sources," Environmental
    Science & Technology, 7, 235.

6. Henry, R. (1977). A Factor Model of Urban Aerosol
    Pollution. Ph.D. Dissertation, Oregon Graduate Center,
    Beaverton, Oregon.

7. Hadley, G. (1962). Linear Programming, Addison Wesley,
    Reading, Massachusetts.

8. Dunker, A.M. (1979). "A Method for Analyzing Data on the
    Elemental Composition of Aerosols," presented to the American
    Chemical Society, Sept. 12, 1979, Washington, D.C.

9. Watson, J.G. (1979). Chemical Element Balance Receptor
    Model Methodology for Assessing the Sources of Fine and Total
    Suspended Particulate Matter in Portland, Oregon, Ph.D.
    Dissertation, Oregon Graduate Center, Beaverton, Oregon.

10. Britt, H. I. and Luecke, R.H. (1973). "The Estimation of
    Parameters in Nonlinear, Implicit Models," Technometrics,
    15, 233.

11. Kleinman, M.T., Pasternack, S., Eisenbud, M. and Kneip, T.J. (1980). "Identifying and Estimating the Relative Importance of Sources of Airborne Particles," Environmental Science & Technology, 14, 62.

12. Hammerle, R.H. and Pierson, W.R. (1975). "Sources and Elemental Composition of Aerosol in Pasadena, California, by Energy-Dispersive X-ray Fluorescence," Environmental Science & Technology, 9, 1058.

13. Neustadter, H. E., Fordyce, J. S. and King, R. B. (1976). "Elemental Composition of Airborne Particulates and Source Identification: Data Analysis Techniques," Journal of the Air Pollution Control Association, 26, 1079.

14. Moyers, J. L., Ranweiler, L. E., Hopf, S. B. and Korte, N. E. (1977). "Evaluation of Particulate Trace Species in Southwest Desert Atmosphere," Environmental Science & Technology, 11, 789.

15. Cahill, T. A., Eldred, R. A., Flocchini, R. G. and Barone, J. (1977). "Statistical Techniques for Handling PIXE Data," Nuclear Instruments and Methods, 142, 259.

16. Hopke, P. K., Gladney, E. S., Gordon, G. E., Zoller, W. H. and Jones, A. G. (1976). "The Use of Multivariate Analysis to Identify Sources of Selected Elements in the Boston Urban Aerosol," Atmospheric Environment, 10, 1015.

17. Alpert, D. J. and Hopke, P. K. (1980). "A Quantitative Determination of Sources in the Boston Urban Aerosol," Atmospheric Environment, 14, 1137.

18. McDonald, G. C. and Schwing, R. C. (1973). "Instabilities of Regression Estimates Relating Air Pollution to Mortality," Technometrics, 15, 463.

19. Henry, R. C. and Hidy, G. M. (1979). "Multivariate Analysis of Particulate Sulfate and Other Air Quality Variables by Principal Components," Atmospheric Environment, 13, 1581.

20. Gaarenstroom, P. D., Perone, S. P. and Moyers, J. L. (1977). "Application of Pattern Recognition and Factor Analysis for Characterization of Atmospheric Particulate Composition in Southwest Desert Atmosphere," Environmental Science & Technology, 11, 796.

21. Box, G. E. P. and Jenkins, G. M. (1976). Time Series Analysis: Forecasting and Control. Holden Day, San Francisco.

22. Bradley, J. V. (1968). Distribution Free Statistical Tests, Prentice-Hall, Inc., Englewood Cliffs, New Jersey.

23. Bradway, R. M., and Record, F. A. (1976). National Assessment of the Urban Particulate Problem, EPA-450/3-76-025, Research Triangle Park, North Carolina.

24. Gartrell, G. and Friedlander, S. K. (1975). "Relating Particulate Pollution to Sources: The 1972 Aerosol Characterization Study," Atmospheric Environment, 9, 279.

25. Pace, T. G. (1979). "An Empirical Approach for Relating Annual TSP Concentrations to Particulate Microinventory Emissions Data and Monitor Siting Characteristics," EPA-450/479-012, Research Triangle Park, North Carolina.

26. Zinmeister, A.R. and Redman. T.C. (1980). "A Time Series Analysis of Aerosol Composition Measurements," Atmospheric Environment, 14, 201.

27. Bevington, P.R. (1969). Data Reduction and Error Analysis for the Physical Sciences. McGraw Hill, New York,

28. Armstrong, J. A., Russell, P. A., and Williams, R. E. (1978). "Balloon-Borne Particulate Sampling for Monitoring Power Plant Emissions," EPA/600/7-78/205, Research Triangle Park, North Carolina.

29. Mroz, E. J. (1976). The Study of the Elemental Composition of Particulate Emissions from an Oil-Fired Power Plant. Ph.D. Dissertation, University of Maryland, College Park, Maryland.

30. Rheingrover, S. W. and Gordon, G. E. (1980). "Identifying Locations of Dominant Point Sources of Elements in Urban Atmospheres from Large, Multi-Element Data Sets," proceedings of the 4th International Conference on Nuclear Methods in Environmental and Energy Research. University of Missouri, Columbia, Missouri.

31. Larkin, R. and Johnson, G. L. (1979). Environmental Assessment Data System User Guide - Draft. EPA/IERL 68-02-2699, Research Triangle Park, North Carolina.

32. Katz, M. (1980). "Advances in the Analysis of Air Contaminants: A Critical Review," Journal of the Air Pollution Control Association, 30, 528.

RECEIVED May 22, 1981.

Air Particulate Control Strategy Development

A New Approach Using Chemical Mass Balance Methods

JOHN E. CORE—Office of Air Quality Planning and Standards,
U.S. Environmental Protection Agency, Research Triangle Park, NC 27711

PATRICK L. HANRAHAN—Oregon Department of Environmental Quality,
P.O. Box 1760, Portland, OR 97207

JOHN A. COOPER—Oregon Graduate Center, 1900 N.W. Walker Road,
Beaverton, OR 97006

Recent advances in source apportionment receptor models
have, for the first time, led to the development of regional
particulate control strategies. Source impacts assigned
using a Chemical Mass Balance (CMB) model, have been used
with dispersion modeling to identify emission inventory
deficiencies and improve modeling assumptions. The
Chemical Mass Balance model is a method of assigning source
impacts given detailed information on the chemical
"fingerprint" of both the ambient particulate and source
emissions within the airshed. Quantitative estimates of
source contribution were identified with relative
uncertainties ranging from $\pm$ 5% to 30%.

Dispersion model source impact estimates, following
comparison to the CMB results, were significantly improved
after emission inventory deficiencies were corrected. Final
modeling results then provided realistic source impact
estimates which could be confidently used for strategy
development.

Presented as an overview of the State of Oregon's
unique approach to particulate control strategy development, this review was prepared to provide those responsible
for airshed management with new information on source impact
assessment methods. (This material is available in the
form of an audio-visual program suitable for presentation
before public, regulatory or private interest groups).

0097-6156/81/0167-0107$05.00/0
© 1981 American Chemical Society

Portland, Eugene, and Medford, Oregon are three cities
which share an air pollution problem common to many other
communities across the country - suspended particulate
air quality violations. Faced with limited airshed capacity,
expanding emission growth and Clean Air Act requirements to
attain particulate air quality standards, the Oregon
Department of Environmental Quality, in 1975, began a five
year program of data collection designed to understand each
community's problem. The purpose of these programs has been
to provide the best technical information possible upon which
control strategies can be built. Results from this work have
played a key role in identifying contributing sources, improving modeling results and in adopting control programs that the
Department and the community can implement with confidence.
     This is a brief review of the Department's approach to
control strategy development. It is unique because it combines
the advantages of two different source apportionment models
to arrive at the source impact information used in strategy
development: a Chemical Mass Balance (CMB) Model to estimate
source impacts using measured ambient particulate composition
data, and traditional dispersion model estimates of impacts.
Although dispersion modeling and Chemical Mass Balance methods
have been used in several source apportionment studies, the
work reported here represents the first attempt to bring the
best features of these techniques together within the framework of a single study (1,2).
     The primary focus here is on work completed in the
Portland Air Quality Maintenance area in Northwest Oregon,
although the improvements to the meteorological data, emission inventories and dispersion model have been completed
in all three cities.

Data Base Improvement Programs
     In 1970, new efforts were underway to solve Portland's
suspended particulate problem. Early efforts relied on available emission factors and industrial source testing, as a
basis for the emission inventory. The inventory was then used,
with a proportional rollback model, as a basis for the new
strategy. New industrial controls were installed which
resulted in a 60,000 tons per year region-wide reduction in
industrial emissions. Although progress toward cleaner air
was made, air quality standard violations caused by then
unknown sources continued and modeling efforts failed to
account for over one-half of the particulate mass.
     A new approach to identifying contributing sources was
badly needed if a new round of emission control regulations
were to be successful. Following technical review of the
alternatives, a comprehensive plan incorporating a Chemical
Mass Balance(3,4) receptor model was adopted. This

technique distinguishes source contributions from one another
by statistically matching the results from extensive chemical
analysis of ambient particulate to those obtained from sources
in the airshed. In Portland, certain sources were easily
identified using "tracer" elements uniquely associated with
one source. Automotive exhaust, for example, is by far the
largest source of lead. Other emissions, such as re-entrained
road dust, must be identified using a number of elements, none
of which are unique to any one source.
    The overall program design consisted of a five-step process
leading to State Implementation Plan revisions:
    1.   Identification of source contributions in the ambient
air using the Chemical Mass Balance. This work was completed as
part of the Portland Aerosol Characterization Study, PACS.
    2.   Dispersion model estimates of source contributions
using source emission and meteorological data for the one year
period of PACS sampling;
    3.   Comparison of the PACS-CMB source impact estimates to
the dispersion model-predicted source impacts;
    4.   Completion of emission inventory and modeling assump-
tion improvements to match dispersion model source impacts to
CMB results; and
    5.   Dispersion modeling of control strategy alternatives.
The first step in the plan was one of source identification
based on air samples.
    After a year of staff design, fund raising and review of
the meteorological, emission inventory and modeling adequacy,
the Portland Aerosol Characterization Study (PACS), Step 1 of
the plan, began. Participation by local government, business
interests and industry was actively sought throughout the PACS
design, sampling and data analysis phases in the hope that all
sectors of the community could gain confidence in the study
results. A public advisory committee was formed to help guide
the project and reveiw early drafts.
    The PACS study was a three year effort designed to iden-
tify major aerosol source types within the Portland Air Quality
Maintenance Area and quantatively determine their contribution
to particulate levels. The PACS represents the first major study
designed from the beginning to provide all of the data required
by the CMB method. Fine and coarse particulate samples from 37
sources, representing 95% of Portland emission inventory were
chemically characterized for 27 chemical species. The same
species were measured on over 2000 individual fine and coarse
ambient particulate samples over a one-year period. Six
sampling locations representing background air quality; residential, commercial and industrial land use area were included
in the study, resulting in over 1700 CMB calculations. The
source contribution estimates derived from the CMB work were
stratified by meteorological regime, season of the year, sampling site and particle size fraction.

Portland Aerosol Characterization Study Results
    Sources of Portland's total suspended particulate mass
(Figure 1) were successfully identified by Chemical Mass
Balance methods (5). The key results of the study were as
follows:
        *Soil and road dust was found to be the largest
        single source, accounting for 55% of the parti-
        culate. Although several minor sources of rural
        dust were included in the area's inventory, the
        study identified a 19,400 ton per year deficiency
        in the paved road dust emission inventory.
        *Vegetative burning was found to account for as
        much as 40% of the total particulate mass on cold
        winter days, or nearly 9% annually. This led to the
        identification of a source of 6500 tons/year of
        previously uninventoried emissions and prompted
        major efforts to reduce impacts from residential
        wood burning.
        *Industrial emissions collectively accounted for
        only 5% of the particulate mass, a result that was not
        surprising given that these sources were well con-
        trolled before field sampling began. Some
        industrial source impacts, however, may not have
        been identifiable because of emissions that chemi-
        cally resembled geologic sources.
        *Secondary particulates were found to account for
        about 8% and automotive exhaust about 10% of the
        annual average mass.
    In all, about 92% of the total suspended particulate mass
was assigned to specific sources or chemical classes with the
remaining 8% likely made up of water, ammonium salts and
other unidentified species. Figure 2 shows the composition
and source contributions to the fine particulate fraction
less than 2.5μm. Figure references to volatilizable and
non-volatilizable carbon are operational definitions (6) for
organic carbon (volatile in an oxygen-free atmosphere at
temperature less than 850°C) and elemental carbon. These
classification differ from vegetative burning in that they
include only carbon whereas burning emissions are about
60-70% carbon. Fine particle sources were also identified
and used during evaluations of control strategy alternatives.
Dispersion Model Estimates of Source Impacts
    Step 2 required identification of source impacts by
airshed modeling. Wind speed, direction, mixing height,
and emission data bases designed to represent conditions on
PACS sampling days were used to insure that the CMB impact
estimates could be directly compared to model predictions
for each sampline site.
    The Portland airshed dispersion model GRID (7) is a con-
servation of mass, advection-diffusion code designed to perform

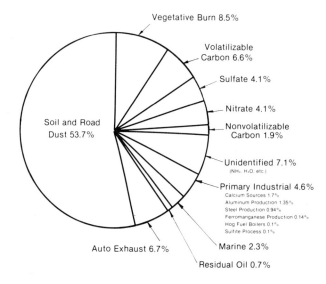

Figure 1.   Sources of total particulate: annual average of the downtown Portland sampling site

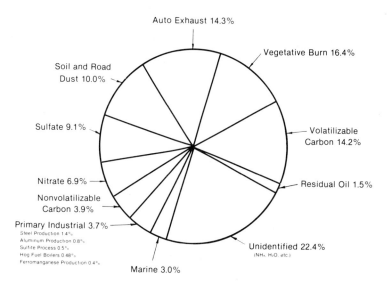

Figure 2.   Sources of fine particulate: annual average of the downtown Portland sampling site

well in the rough terrain that characterizes the Portland area. Required inputs for each of the 5000, 2km grid cells include topography and wind flow fields by hour for each of eight meteorological regimes into which the annual weather patterns were classified. Spatially and seasonally resolved point and area source emissions and stack parameters were also developed.

Three source classes were selected for comparison of CMB and model results. Two were of an area source nature (road dust and automotive exhaust) and one point source category (residual oil combustion). These sources were selected for comparison for the following reasons:

*Dust from paved and unpaved roads is the most predominant source in Portland. Accurate emissions data is critical to the model's ability to account for all of the measured mass, as well as to the development of a successful strategy. Although the CMB model cannot distinguish between dust emissions from paved and unpaved roads, it can typically assign total dust impacts to within a 5-6% uncertainty.

*Leaded automotive tailpipe exhaust was selected because it can be accurately estimated by CMB and is the third most abundant contributor to the annual particulate mass, after road dust and vegetative burning.

*Residual oil combustion impact served as a test of the model's ability to predict point source emissions. Since vanadium and nickel emissions in Portland are almost totally associated with residual oil use, the CMB method was able to assign impacts with a high degree of confidence by using these two elements as chemical tracers.

Model predictions for these sources were prepared for each of eight meteorological regimes, appropriately weighted and combined to provide annual impact estimates for each grid and each PACS monitoring site.

Comparison of CMB and Dispersion Model Impact Estimates

Step 3 was a comparison of the CMB and GRID model impact estimates. CMB-estimated source contributions to the background aerosol immediately upwind of the study area were subtracted from those obtained from the urban site data to provide a "local CMB" value against which the GRID dispersion model could be compared. Data collected during the North-wind Regime was used during this work since computer simulation costs were only one-eighth of that required for an annual simulation. Initial source apportionment estimates from these two models were as follows:

*Automotive exhaust comparisons (shown in Figure 3) at three sites were well within the CMB estimated uncertainty. One of the sites, however, appeared to be underpredicted. Although

a closer examination of the emission inventory
near Site 5 was needed, the initial CMB-model
impact comparisons were encouraging. This initial
comparison indicated that the Portland GRID cell
dispersion model was able to simulate automotive
exhaust impacts and area source emissions reasonably
well.
*Road dust comparisons shown in Figure 4 were poor
with major underpredictions at all locations by the
GRID model. More work was obviously needed before
it could realistically predict dust impacts.
*Residual oil impact estimates by modeling provided
a severe test of GRID's capacity since the CMB
impact estimates were small (less than one-quarter
$\mu g/m^3$) and the physical basis of the model inherently limits it's ability to predict point source
plume transport. Since initial comparisons
(Figure 5) showed GRID estimated impacts to be over-
predicted at all sites relative to CMB estimates,
further improvements to the data base were suggested.

Overall, annual model verification results for all
sources were relatively poor with the dispersion model predictions consistently underestimating both the CMB-derived
estimates and the measured TSP mass data.

Model and Emission Inventory Improvements

The CMB-dispersion model comparisons had identified data
base deficiencies that would not have been apparent had the
model predictions been compared only to measured particulate
mass. Correction of the emission inventory and model inade-
quacies was, however, critical to improving the model's
source impact predictions and to the strategy's success.
Step 4, then, involved nearly six months of research into
the emission inventory, modeling and meteorological assumptions used in the modeling. The results were as follows:

Auto exhaust emission calculation errors were found
which tripled tail pipe emissions near Site 5
and increased the model predicted impact by nearly
10%. The error was caused when emissions from a
heavily traveled road near the grid boundary was
incorrectly assigned to a neighboring grid. This
helped to explain some of the model underpredictions.
Residual Oil impact overpredictions were traced to
three errors:
*The emissions were incorrectly assumed to be con-
stant throughout the year. Corrections included
modification of the dispersion model code to
accomodate specific monthly operating schedules for
each meteorological regime.
*Topographical data within critical cells were
reviewed. Since the dispersion model can only

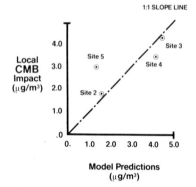

Figure 3. Motor vehicle tailpipe emissions: CMB vs. Model (annual simulation)

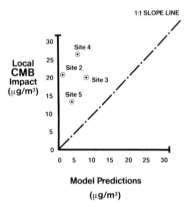

Figure 4. Road dust impacts—north wind regime (initial predictions)

accept a single value per cell as the ground elevation, errors caused by topographical variations
within a cell were introduced. The model consists
of five cells in the vertical direction to accomodate
vertical differences in wind speed and direction.
Because the downwind plume impact is assigned to one
of the five vertical cells, relatively minor changes
in plume height can result in major changes in predicted concentrations at the receptor if the change
moves the plume into the ground level cell. To
correct the errors, stack heights in each cell had
to be adjusted to reflect differences between the
true stack and receptor heights.
*Large impacts from one of the airshed's major
sources was traced to an unrealistic operating
schedule.
Following corrections, residual oil impacts agreed quite well
with CMB estimates.

Road dust underpredictions, based on the initial EPA
generalized paved road dust factor, were the most
serious problem. Since there are very few unpaved
roads within the AQMA, attention turned to upward
adjustment of the paved road dust emission factor.
A modified emission factor (EF) was developed based
on an assumed consistent relationship between particulate tailpipe and road dust impact:

$$\frac{EF_{tailpipe\ emissions}}{EF_{paved\ road\ dust}} = \frac{CMB_{tailpipe\ impact}}{CMB_{road\ dust\ impact}}$$

Further seasonal adjustments in the new factor were
made to account for rainfall.
Finally, data from studies in Seattle, Washington (8)
suggested a 10 to 20 fold increase in street dust emissions
in heavy industrial areas as compared to commercial, land
use areas. Adjustments were made to each grid's emissions
based on information from land use maps and detailed data on
unpaved roads.
The final road dust inventory was increased by 19,400 tons
per year, a 600% increase in road dust emissions. Even so, the
overall AQMA emission factor of 3.2 grams per vehicle mile is
less than the 5.6 grams per vehicle mile currently recommended
by EPA. Modeled street dust impacts now agreed with CMB
results shown in Figure 6. As a result of these adjustments,
the updated emission inventory was increased substantially
(see Figure 7). Road dust is now estimated to be over one-half
of the total emissions. As a result of the PACS program,
emissions from residential wood space heating were added as an
important new source category. This latter category is pro-
jected to be the fastest growing emission source in the near
future.

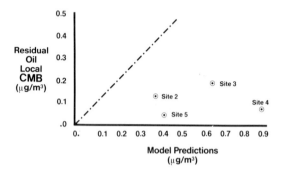

Figure 5. Residual oil impacts—north wind regime (initial predictions)

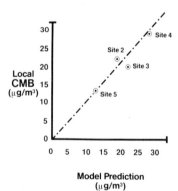

Figure 6. Road dust impacts—north wind regime (corrected inventory)

Final GRID dispersion model results were significantly improved and systematic underpredictions eliminated. Measured and model-predicted background values agreed more closely than before and the community could have greater confidence in the dispersion model's ability to evaluate the effectiveness of alternative control strategy. Figure 8 illustrates the improvements in model verification following emission inventory and modeling assumption improvements.

Control Strategy Effectiveness

Using the validated dispersion model as the key tool in the strategy development, efforts turned to Step 5, Control Strategy Development. Table I presents predictions of 1987 particulate air quality based on future year emission data bases, and incorporates all of the identified inventory corrections, and emission growth projections.

The 1977 to '87 growth increment was calculated from the modeling and added to the measured particulate values to arrive at a 1987 design value upon which annual standard attainment could be judged. Twenty-four hour attainment status was similarily determined, but was based on a worst case meteorological regime and emissions typical of those occurring on violation days. Results required a 2 to 24 $\mu g/m^3$ annual average reduction at various sites to achieve the 60 $\mu g/m^3$ secondary TSP standard and a 19 to 104 $\mu g/m^3$ improvement in 24-hour particulate air quality to achieve the 150 $\mu g/m^3$ secondary standard.

Model predictions of numerous point and area source strategies were completed, an economic analysis of the alternatives was developed, and the results placed before a public advisory committee. The cost-effectiveness and fine particle benefits identified for each strategy have provided the public with a sound basis for their recommendations. Based on an early staff analysis, the most cost effective alternatives included:

> *Reductions in vehicle traffic in violation areas at a net annual cost savings of about $100,000 per $\mu g/m^3$. Associated benefits include energy savings, fewer road and auto maintenance costs and important carbon monoxide and hydrocarbon emission benefits.
> *Cleanup of winter road sanding dust at a cost of $1700 to $8300 per $\mu g/m^3$.
> *Construction site trackout control in particulate standard violation areas estimated to cost about $52,000 per $\mu g/m^3$.

Less cost effective controls include:

> *Further controls on local point sources at $2.6 million per $\mu g/m^3$.
> *Adoption of a 0.5% sulfur content limit for residual oil at $4.2 million per $\mu g/m^3$.

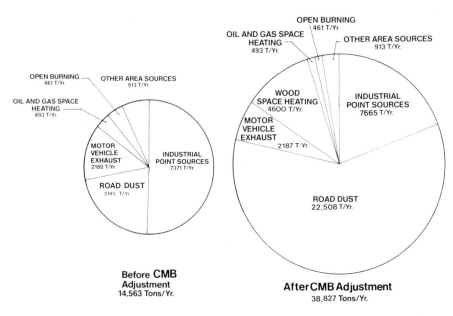

Figure 7.   Portland AQMA emission inventory—1977 total particulate

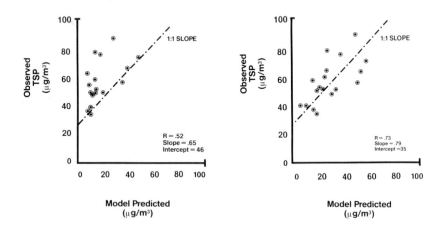

*Figure 8.   Portland AQMA annual model predictions before (left) and after (right)
emission inventory improvements*

Table I
Particulate Control Alternatives
Model Predicted Source Impacts for 1987
($\mu g/m^3$, 24 Hour Worst Case Meteorology)

Site	Design Values* 1977	Design Values* 1987	Reduction Needed To Meet Secondary Standard in 1987	Road Dust Impact	Auto Exhaust	Wood Burning	All Point Sources
1.	197	219	69	106	1.0	8.2	7.1
2.	161	169	19	82	4.2	6.9	6.2
3.	223	254	104	120	5.2	37.0	7.4

*Predicted particulate concentrations on 2nd highest day.

Although the control of fugative emissions appears
to be the most cost effective approach to particulate standard
attainment, few benefits in regional visibility improvement
and public health would be expected from such a strategy.
A balance between control of fugative dust emissions and
sources of fine particulates is perhaps the most appropriate
approach to standard attainment.

Strategy Development in Other Airsheds

Source impact studies using the Chemical Mass Balance-
dispersion modeling approach have also been completed in
Eugene and Medford, Oregon(9,10). As in Portland, the work
has identified emission inventory and modeling improvements
which, when corrected, have greatly improved modeling results.
Figure 9 shows dispersion model results for the Medford airshed before and after emission inventory and model improvements. Several of the most significant results of this work
are:

*Road dust emission inventory errors were
corrected after comparing dispersion model
and CMB results. Road dust emissions were
increased by 2300 tons in Medford as a
result of emission factor adjustments and
changes in the unpaved road dust inventory.
In Eugene, new measurements of traffic volume and vehicle speeds on unpaved roads led
to a 2600 ton per year reduction in road
dust emissions.

*Residential wood burning emissions,
identified by a telephone survey of home-
owners, were verified by comparing
dispersion model and CMB vegetative burning estimates.

*Point source plume trapping assumptions
used in the Eugene airshed modeling were
verified by comparing alternative modeling
assumptions of sulfate emission impacts to
measured sulfate levels.

*Major changes in the Medford model's
treatment of area source impacts were completed after early results consistently
underpredicted road dust impacts measured
by CMB. As a result, model performance
improved dramatically.

Control Strategy Tracking Using CMB Methods

Following completion of the data base improvement programs, a better, yet still incomplete, knowledge of the
sources of the particulate is being used to direct control
programs aimed at achieving air standards with minimum cost
to the community. Had the modeling work not been compared to
the CMB results, future controls would likely have been
directed toward traditional, industrial sources.

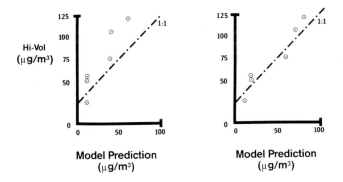

Figure 9. *Medford annual model predictions—1978 (arithmetic mean). Hi-vol vs. model predicted TSP mass before (*left*) and after (*right*) model improvements*

Without the benefit of the data base improvement programs, $27
million dollars in industrial controls would likely have been
the key element in a new control strategy yet would have only
provided one-third of the reduction required to meet standards.
Instead, a more effective mix of point and area source controls
will likely be adopted and, at considerably less cost to (and
with considerably more effectiveness for) the community. Once
the capability is developed, CMB estimates of source impacts
can be used to track the progress of the strategy on a source-
class basis using actual air samples, (in addition to emission
estimates) to measure the effectiveness of the controls.

Oregon's Source Apportionment Program and the Future

Source apportionment programs using Chemical Mass Balance-
dispersion modeling, optical microscopy, and other methods has
become, and will continue to be, an important part of the
Oregon's particulate air quality management program. During
the past two years, the Department's laboratory division has
acquired all the necessary analytical and air monitoring cap-
abilities needed to obtain Chemical Mass Balance input data
and interactive CMB software is available to Air Quality
Division staff. Information is also being gathered in several
Oregon communities to define source contributions from wood
combustion, to resolve impacts currently listed as "unidentified" and improve source characterization data. These programs
are providing information on current source impacts, as well as
an independent means of tracking the success of control
programs.

A historical data base of size-resolved particulate data
suitable for source apportionment studies provides a wealth of
data useful to regulatory authorities. As new programs begin
to focus on inhalable and respirable particulate standards,
this data will become increasingly valuable.

Conclusions

The development of new source apportionment methods have,
for the first time, led to the development of regional particulate control strategies. Source impacts assigned using a
chemical mass balance (CMB) model have been used in association with airshed dispersion models to identify emission
inventory deficiencies and improve modeling assumptions.

Experience gained during the past four years has shown
that data base improvement programs such as those implemented
in Oregon have provided new perspectives into the nature of
each airshed's emissions. Subsequent major changes in each
community's emission inventories have improved dispersion
modeling results and provided a level of dispersion model
verification impossible to attain using traditional hi-vol
measurements alone.

Although the current technology of receptor models is
still in an early stage of development, the rapid growth of
these techniques is certain to provide an important new tool
to regulatory agencies within the immediate future.

Literature Cited

1. U.S. Environmental Protection Agency "Guidelines on Air Quality Models", EPA-450/2-78-027, April 1978.

2. Cooper, J.A.; Watson, J.G. "Receptor Oriented Methods of Air Particulate Source Apportionment" J. Air Pollution Control Association, 1978, 30 (10).

3. Friedlander, S.K. "Chemical Element Balance and Identification of Air Pollution Sources" Env. Sci. & Technol., 1973, 7 (3).

4. Watson, J.G. "Chemical Element Balance Receptor Model Methodology For Sources of Fine and Total Suspended Particulate Matter in Portland, Oregon." Oregon Graduate Center, February 1979.

5. Cooper, J.A.; Watson, J.G. "Portland Aerosol Characterization Study", Final Report to the State of Oregon Department of Environmental Quality, July 1979.

6. Johnson, R.L.; Huntzicker, J.J. "Analysis of Volatilizable and Elemental Carbon in Ambient Aerosols", Conference on Carbonaceous Particules in the Atmosphere, Berkeley, CA, 1978.

7. Fabrick, A.J.; Sklarew, R.C. "Oregon/Washington Diffusion Modeling Study" Xonics, Inc., 1975.

8. Roberts, J.; Watters, H.; Austin, F.; Crooks, M. "Particulate Emissions For Paved Roads in Seattle and Tacoma Non-Attainment Areas", Puget Sound Air Pollution Control Agency, July 1979.

9. Core, J.E.; Greene, W.T.; Terraglio, F.P. "Slash Burning Particulate Impact Analysis in Oregon's Williamette Valley" State of Oregon Department of Environmental Quality, June 1979.

10. Cooper, J.A. "Medford Aerosol Characterization Study - Application of Chemical Mass Balance Methods to the Identification of Major Aerosol Sources in the Medford Airshed", Interim Report to the State of Oregon Department of Environmental Quality, November 1979.

RECEIVED March 25, 1981.

Chemical Species Contributions to Light Scattering by Aerosols at a Remote Arid Site

Comparison of Statistical and Theoretical Results

J. R. OUIMETTE[1] and R. C. FLAGAN

California Institute of Technology, Pasadena, CA 91125

A. R. KELSO

Naval Weapons Center, China Lake, CA 93555

A one-year, aerosol-light extinction experiment was conducted at China Lake, California, an arid site in the Mohave Desert. Measurements of the aerosol size distribution, the light scattering coefficient, and fine aerosol chemical composition were made. Multiple regression analysis was applied to the measured particle light scattering coefficient and fine aerosol species mass concentrations from 61 filter samples collected during 1979. Contributions of various aerosol species to the particle light scattering coefficient, b_{sp}, were estimated. The statistically estimated contributions were compared with those determined theoretically using measured aerosol mass distributions. It was found that the statistically inferred species contributions to b_{sp} agreed qualitatively with those calculated theoretically using measured aerosol distributions. Regression analysis overestimated the contribution of sulfate relative to that calculated theoretically. Using measured 1979 values, a light extinction budget was calculated for China Lake. Measured mass extinction coefficients were used to predict the reduction in visibility at China Lake which would occur by increasing the concentrations of various aerosol species.

A proper control strategy for visibility protection at a given location requires information about the relationship between each source and its contribution to a quantity called the extinction coefficient, b_e. The visual range, V_R, and b_e are related through the Koschmieder equation ([1]).

$$V_R = 3.9/b_e$$

---

[1] Current address: Chevron Research Company, Richmond, CA 94802

0097-6156/81/0167-0125$08.00/0

The extinction coefficient can be expressed as the sum of terms which account for light scattering and absorption by gases and suspended particles ($\underline{2}$):

$$b_e = b_{sg} + b_{ag} + b_{sp} + b_{ap}$$

where a, s, g, and p denote absorption, scattering, gases, and particles, respectively. The Rayleigh scattering coefficient, b_{sg}, has a value integrated over the solar spectrum of $15 \times 10^{-6} m^{-1}$ at sea level and 25°C and varies linearly with air density. It provides an upper bound on visual range of about 260 km. It is not unusual in many remote locations in the western United States for the visibility to be limited by Rayleigh scattering due to air molecules ($\underline{3}$). However, in many rural and urban locations, b_e is dominated by light scattering due to particles, b_{sp}. To assess the contribution of a particular source to visibility reduction at a given location, it is necessary to determine its contribution to b_{sp}, b_{ap}, and b_{ag}. As a good approximation, the contribution of a particular source to b_{ag} would be directly proportional to its contribution to the total NO_2 concentration. The contribution of the source to the scattering and absorption of light by aerosols, b_{sp} and b_{ap}, is not so straightforward. One must first link the source to concentrations of each chemical species in the aerosol. Second, the contribution of each chemical species to the particle scattering and absorption coefficients must be determined.

The contributions of aerosol chemical species to the extinction coefficient can be estimated from knowledge of their mass distributions, densities, and refractive indices. It is assumed that the particles can be represented as spheres. For a mixture in which the composition is a function of particle size and all particles of a given size have the same composition, defined here as a specific mixture, the contribution of species i becomes ($\underline{4}$):

$$b_{epi} = \int_0^\infty E_e(m_a(x),x,\lambda)f_i(x)dx \qquad (1)$$

where the mass distribution of species i is $f_i(x) = dM_i/dx$, $x = \log(D/D_0)$, E_e is the mass extinction efficiency for species i, and $m_a(x)$ is the volume average refractive index as a function of x. Strict application of Equation (1) requires information on all species.

The species mass extinction efficiency, E_e, can be theoretically determined from Mie's classical solution to light extinction by a sphere in an infinite medium. Computer routines are available to calculate single particle extinction efficiencies, and, hence, E_e ($\underline{5}$). If the mass distribution of each species is experimentally measured, then the optical property, b_{epi}, can be theoretically predicted using Equation (1) and calculated extinction efficiencies. It is, however, not presently possibly to measure directly the effect of a particular aerosol chemical species in a multicomponent aerosol on the extinction coefficient.

In a separate study (6) aerosol species mass distributions were successfully used to calculate the contribution of each species to the extinction coefficient. Unfortunately, such detailed data is not usually available. At most air monitoring stations, only the total aerosol species mass concentrations, M_i, are determined from filter samples. Statistical methods have been used to infer chemical species contributions to the particle light extinction coefficient (3). For such analyses it is assumed that b_{ep} can be represented as a linear combination of the total species mass concentrations, M_i, viz.,

$$b_{ep} = \sum_{\alpha=1}^{N} \alpha_i M_i \qquad (2)$$

Ouimette and Flagan (4) have shown that this assumption is valid for a specific mixture provided:

a) The normalized mass distribution for each species does not vary.

b) The normalized total aerosol volume distribution is preserved.

c) The refractive indices of all chemical species are equal.

d) The partial molar volume of each species must remain constant with the addition or removal of other species, i.e., aqueous solutions would not be permitted.

For an external mixture of pure, single component aerosol particles, only the first condition, (a), is necessary.

Requirement (c) will not be met for soot which contributes to extinction primarily by absorption. For this reason, a satisfactory linear relationship is more likely to be found for aged aerosols (particle compositions mixed by coagulation) only between scattering due to particles and the mass concentrations:

$$b_{sp} = \sum_{i=1}^{N} \alpha_i N_i \qquad (3)$$

These limitations on the statistical inference of species contributions to the particle light extinction coefficient raise several questions: (i) Is the statistical analysis of filter data a valid method for determining species contributions to the extinction coefficient of atmospheric aerosols? (ii) Can we place confidence limits on the quality of the statistical results? and (iii) Can the quality of the statistical results be enhanced through improved sampling techniques?

To test the applicability of statistical techniques for determination of the species contributions to the scattering coefficient, a one-year study was conducted in 1979 at China Lake, California. Filter samples of aerosol particles smaller than 2 μm aerodynamic diameter were analyzed for total fine mass, major chemical species, and the time average particle absorption coefficient, b_{ap}. At the same time and location, b_{sp} was measured with a sensitive nephelometer. A total of 61 samples were analyzed. Multiple regression analysis was applied to the average particle scattering coefficient and mass concentrations for each filter sample to estimate α_i and each species contribution to light scattering, b_{spi}. Supplementary measurements of the chemical-size distribution were used for theoretical estimates of each b_{spi} as a test of the effectiveness of the statistical approach.

EXPERIMENTAL TECHNIQUE

Location

Aerosol sampling was routinely conducted from February through December, 1979, at China Lake, California, an arid site with no significant local sources of air pollution (7). Shown in Figure 1, the sampling site is located in the Mohave Desert within the U.S. Naval Weapons Center (NWC), a missile research and testing facility. The site is about 200 km northeast of Los Angeles in extreme northeastern Kern County. It is 100 feet from a paved access road which averages about 30 vehicles per day. The sampling site is at an elevation of 670 m above mean sea level. China Lake is characterized by abundant sunlight, low relative humidity, and hot daytime temperatures in the summer.

The particle light scattering coefficient has been continuously measured at this location since 1976. Measurements of the particle size distribution have been made daily since 1978, providing the data base necessary to assess the variability of the normalized aerosol volume distribution.

Aerosol Measurement and Sample Collection

Aerosol sampling instruments were housed in an air conditioned monitoring trailer operated by NWC. The particle light scattering coefficient, b_{sp}, was continuously measured with modified MRI Model 1561 integrating nephelometer. The nephelometer automatically zeroed daily and was calibrated quarterly. The estimated error in measurement was $\pm 2.5 \times 10^{-6} m^{-1}$ over the range $0-100 \times 10^{-6} m^{-1}$.

The number distribution of particles having diameters from 0.03-10 um was measured twice daily on most weekdays with a Thermo Systems electrical aerosol analyzer (EAA) (8) and a Royco Model 202 Optical Particle Counter (OPC). These measurements were performed at a site three miles southeast of the air

monitoring trailer (9). The constants of Whitby and Cantrell
(10) were used to convert current changes in the EAA to particle
number concentrations. In the region between 0.3 and 0.6 μm,
where both instruments provide data, the particle number concentration determined with the EAA was generally greater than that
determined with the OPC.

Aerosol for chemical analysis was sampled in the air monitoring trailer through a 1.3 cm ID stainless steel pipe. The
air inlet was about 1 m above the roof of the trailer, a total
of 4 m above the ground. Loss of 0.1 μm diameter particles to
the walls due to turbulent diffusion was calculated to be less
than 1% using the method of Friedlander (11). A cyclone preseparator (12) was used to separate the coarse (D > 2 μm) aerosol
from the airstream so that only the fine (D < 2 μm) aerosol
would be collected for analysis. The cyclone was operated at
26-30 liters per minute (1pm) and was cleaned every 8-10 weeks.

The fine particle airstream from the cyclone was sampled by
two total filters in parallel. A Millipore Fluoropore 47 mm
diameter Teflon filter with a 1 μm pore size was used for the
first seven samples. Subsequent samples were obtained with a
0.4 μm pore size 47 mm Nuclepore polycarbonate filter because
particle absorption measurements and elemental analysis by
particle induced X-ray emission (PIXE) were easier and more
accurate using the Nuclepore filters. In parallel with the
Nuclepore filter, a TWOMASS tape sampler collected aerosol using
a Pallflex Tissuequartz tape. The aerosol deposit area was
9.62 cm^2 on the Nuclepore and Millipore filters and 0.317 cm^2 on
the Tissuequartz tape. The flow rate was 16-20 1pm through the
Nuclepore and Millipore filters and 10 1pm through the Tissuequartz tape. Each Millipore or Nuclepore filter was placed in
a labeled plastic container immediately after collected, sealed
with Parafilm, enclosed in a ziplock bag, and placed in a
refrigerator in the trailer. The tape in the TWOMASS sampler
was advanced between samples. The tape sample was removed about
once every 8-10 weeks and stored similarly to the Nuclepore
filters. The TWOMASS was cleaned at that time. All samples
were stored in an ice chest during the return trip to Caltech.
Field blanks were handled identically to the samples. Of approximately 100 filter samples collected in 1979, 61 were selected for
analysis. The 61 were chosen to span the variation in b_{sp} and
to obtain representative seasonal and diurnal samples. Sample
times varied from 6 to 72 hours, with an average of 20.1 hours.

Aerosol was occasionally sampled with low pressure impactors
(LPI's) to obtain sulfur and elemental mass distributions using
the techniques described by Ouimette (13).

Sample Analysis

The aerosol mass on each Nuclepore filter sample was determined gravimetrically. Field blanks were obtained at Zilnez Mesa
by drawing 50 ℓ of filtered air through preweighed Nuclepore

filters. The average aerosol mass on the 5 field blanks was
2.2 µg, with a standard deviation of 7.6 µg. After the mass
determination, each of the filters was cut into four pieces for
additional analyses. All field blanks and field samples were
handled identically for chemical analysis.

Elemental mass concentration - One-third of each Nuclepore
filter was sent to Crocker Nuclear Laboratory, University of
California, Davis, for elemental analysis by particle induced
X-ray emission (PIXE)(14). Masses of many elements from Al to
Pb were determined with this technique, including Si, S, K, Ca,
Fe, and trace species such as V, Ni, and Zn. Detection limits
range from about 20 to 200 ng/cm^2 area density on each filter,
corresponding to about 0.2 to 2 µg/filter. Typical uncertainties in the mass determination of a particular element were
±15-20% or ±2 µg/filter, whichever was larger.

Elemental mass distribution - The aerosol sampled by the
LPI for elemental analysis was impacted on coated mylar films
affixed to 25 mm glass discs. The mylar had been coated with
Apiezon L vacuum grease to prevent particle bound. The LPI
samples were sent to Crocker Nuclear Laboratory for elemental
analysis by PIXE using a focused alpha particle beam of 3 to 4 mm
diameter. Nanogram sensitivities for most elements were achieved
with the focused beam. A detailed description of the PIXE
focused beam technique applied to LPI samples can be found in
Ouimette (13). Based upon repeated measurements of field
samples, the estimated measurement error was about ±15-20%
or twice the minimum detection limit, whichever was larger.

Sulfur mass distribution - The aerosol sampled by the LPI
for sulfur analysis was impacted on vaseline-coated stainless
steel strips backed by 25 mm glass discs. The sulfur mass
deposited on each stage was determined by the technique of flash
volatilization and flame photometric detection (FVFPD)(15).

$SO_4^=$, NO_3^-, NH_4^+ mass concentration - One-half of each Nuclepore filter was analyzed by Environmental Research and Technology,
Inc. (ERT), Westlake Village, California. Their laboratory
determined the masses of aerosol sulfate and nitrate on each
filter by liquid ion chromatography and ammonium by colorimetry.
Based on duplicate analysis of samples and standards the uncertainty in the various determinations per filter were:
±1.2 µg NO_3^-, ±2.2 µg $SO_4^=$, and ±0.3 µg NH_4^+.

Particle absorption coefficient - The time average particle
absorption coefficient was measured from one-sixth of each
Nuclepore filter sample using the opal glass integrating plate
technique (16,13). The filter was held against a piece of
opal glass and illuminated by a narrow beam of light at
0.6328 µm wavelength which had propagated through a 0.25 mm
diameter optical fiber. The following relation (16) was used to
calculate the time average particle absorption coefficient, b_{ap}:

$$b_{ap} = A/S \ln(I_0/I) \hspace{3cm} (4)$$

where A is the areosol deposit area on the filter and S is the total volume of air sampled through the filter. I_o is the light intensity measured through the filter blank, and I is the intensity measured through the filter sample. By varying the filter position with respect to the beam, it was determined that the aerosol was deposited uniformly on the filter.

Carbon - The aerosol collected on the Tissuequartz filters was analyzed for total carbon by the nondestructive technique of proton-induced gamma-ray emissions (17) and graphitic carbon (soot) by calibrated optical reflectance at Washington University, St. Louis, Missouri. The estimated error in measurement of total carbon per filter was ±5 µg or ±20%, whichever was greater. The estimated error for soot was ±1 µg or ±20%, whichever was greater.

EXPERIMENTAL RESULTS

The particle scattering coefficient, b_{sp}, was measured continuously throughout 1979. The monthly hourly average b_{sp} are summarized in Table 1. It is seen that both seasonal and diurnal variations in b_{sp} occurred at China Lake. Winter had the best visibility, with summer the worst. Nighttime b_{sp} values were higher than those in the afternoon; the average ratio of 3-4 AM b_{sp} to 3-4 PM b_{sp} was 1.77. It is likely that the diurnal variation in b_{sp} at China Lake is coupled to the diurnal cycle of afternoon winds and nighttime inversions. Polluted air from urban sources is evidently transported by winds in the early evening and trapped by a stable inversion through the night, causing elevated values of b_{sp}. The inversion lifts the next morning, reducing the b_{sp} by mixing with cleaner air aloft. A minimum in b_{sp} then occurs in the midafternoon before the cycle repeats in the evening. A detailed study of aerosol transport from the San Joaquin Valley and Los Angeles Basin, and its effect on visibility at China Lake, can be found in Reible et al. (18).

A total of 254 particle size distributions were measured throughout 1979. The average normalized volume distribution is plotted in Figure 2. The error bars are standard deviations. It is seen that the distribution is bimodal, with the coarse mode dominating the aerosol volume concentrations. The 1979 average volume concentration of aerosol less than 10 µm diameter was 32.4 $\mu m^3/cm^3$. From its large standard deviation, it is clear that the coarse particle mode exhibited considerable variation throughout the year. Records show that high coarse mode volume concentrations accompanied moderate-to-high wind speeds. The coarse material was very likely wind-blown dust of crustal composition.

The aerosol scattering coefficient distribution was calculated from the aerosol volume distribution, using the method described in Friedlander (11). The resultant distribution is plotted in Figure 3. The contribution of the fine aerosol to visibility degradation at China Lake is seen in this figure.

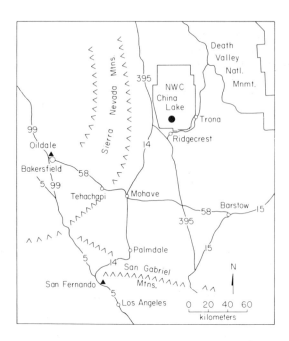

Figure 1.   Location of sampling site within the Naval Weapons Center, China Lake, CA. Site is located in Indian Wells Valley about 6 miles north of Ridgecrest, population (1979 estimate) 16,000. Highway numbers are designated on map.

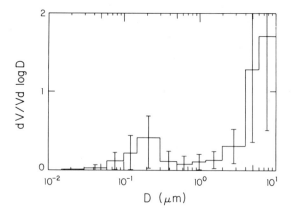

Figure 2.   Normalized aerosol volume distribution, China Lake, CA (1979 average)—average of 254 measurements. The error bars are standard deviations. The distribution is normalized with respect to total aerosol volume concentration of particles less than 10 μm in diameter.

Table 1

Hourly Average Measured Particle Scattering Coefficient, b_{sp}, by Month for China Lake, 1979

Hour, PST	Hourly Average b_{sp}, $10^{-6}m^{-1}$												Ave. 1979
	J	F	M	A	M	J	J	A	S	O	N	D	
0	25	16	18	23	31	34	39	33	35	40	24	28	29
1	24	16	29	22	31	33	39	33	37	37	26	25	29
2	22	16	21	24	31	33	39	34	36	36	25	24	28
3	20	15	23	25	34	35	40	34	36	38	21	23	29
4	22	15	21	22	34	38	41	36	35	36	19	21	28
5	21	15	20	22	34	33	43	39	34	36	19	20	28
6	21	15	20	22	34	33	45	40	33	40	18	19	28
7	22	16	19	23	31	33	44	40	34	40	18	19	28
8	22	16	20	22	30	30	44	38	33	44	19	18	28
9	20	18	17	20	28	27	41	34	32	45	20	19	27
10	18	19	15	20	26	22	33	30	28	38	19	19	24
11	23	16	28	18	23	19	27	25	25	33	17	28	21
12	22	15	13	17	24	29	22	22	25	27	14	17	19
13	19	11	23	17	22	16	20	20	23	24	12	14	18
14	17	11	10	16	21	15	17	19	22	23	10	13	16
15	15	14	9	15	22	14	18	19	21	25	12	15	17
16	14	16	12	15	21	15	20	20	23	21	14	16	17
17	21	15	14	16	22	19	23	21	24	22	17	23	20
18	21	16	17	18	23	21	25	23	26	26	17	28	22
19	23	14	21	22	26	27	31	28	31	32	30	41	27
20	24	21	21	22	27	29	34	30	34	38	35	45	30
21	28	20	21	23	28	32	37	32	35	39	29	47	31
22	23	18	20	24	29	33	38	33	36	42	28	42	31
23	23	17	22	23	31	34	38	33	35	41	24	33	30

1979 Average 25.2

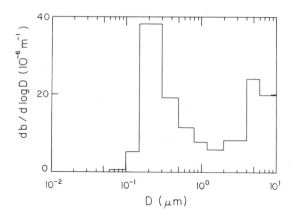

Figure 3. Particle scattering coefficient, China Lake, CA (1979 average). The distribution is calculated from 254 measured aerosol volume distributions assuming m = 1.54–0.0151.

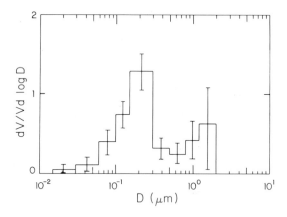

Figure 4. Normalized fine aerosol volume distribution, China Lake, CA (1979 average)—average of 254 measurements. The error bars are standard deviations. The distribution is normalized with respect to total aerosol volume concentration of particles less than 2 μm in diameter.

Over half of the computed particle scattering coefficient is due to particles less than 0.5 μm diameter, although the fine particles contributed an average of only 13% to the aerosol volume concentration. The computed scattering coefficient for each size distribution measurement was compared to the hourly average scattering coefficient measured during the same hour. The results are shown in Table 2. The calculated scattering coefficient is, on the average, 37% larger than the measured value. The difference results from the low sensitivity of the nephelometer to scattering by large particles (19). One may infer that approximately 27% of the scattering is due to coarse particles. When only the fine aerosol (D < 2 μm) is considered, it is seen that the calculated scattering coefficient, b_{spl}, agrees very well with the data. In a statistical analysis, b_{sp} was very well correlated with the volume of fine particles, r = 0.81, but not with the total volume of particles smaller than 10 μm, r = 0.44. Similarly, Trijonis and Yuan (3) found that the total concentration was a poor indicator of visual range in Phoenix.

Because the fine aerosol was found to be responsible for the bulk of light scattering at China Lake, this mode was examined to see if its normalized distribution remained constant throughout 1979. Figure 4 shows the 1979 average aerosol volume distribution at China Lake normalized with respect to the total volume of particles smaller than 2 μm. The error bars represent standard deviations in the 254 measurements. The particle volume distribution in the fine mode is seen to preserve its shape rather well. Over half the fine particle volume is due to particles of less than 0.3 μm diameter.

Aerosol Composition

Table 3 summarizes the 1979 annual average particle extinction coefficient and the mass concentrations of the fine aerosol chemical species estimated by statistical analysis of the 61 filter samples. Organics and sulfates dominated the chemically determined fine aerosol mass at China Lake in 1979. A linear least squares fit between molar concentrations of NH_4^+ and $SO_4^=$ gave a zero intercept, a slope of 1.87 and a correlation coefficient of 0.98. It is therefore assumed that the fine sulfate aerosol was in the form of ammonium sulfate. The mass concentration of carbonaceous and sulfate aerosols were, on the average, comparable in magnitude.

The average and standard deviation of the particle scattering coefficient, b_{sp}, during filter sample collection was $25.4 \pm 13.2 \times 10^{-6} m^{-1}$. This compares favorably with the annual measured average of $25.2 \times 10^{-6} m^{-1}$, indicating that the filter samples were representative of the variation in b_{sp} throughout the year. The average and standard deviation of the particle absorption coefficient, b_{ap}, by filter analysis was $5.1 \pm 2.3 \times 10^{-6} m^{-1}$. On the average, the ratio of the particle absorption to scattering coefficient was 0.22 ± 0.13.

Table 2

Summary of Particle Scattering Coefficient Linear Regression
Relationships, China Lake, 1979

y	x	y_0	m	r
b_{sp} (calc.)	b_{sp} (meas.)	2.48	1.37	0.78
b_{sp1} (calc.)	b_{sp} (meas.)	0.5	0.92	0.88
b_{sp} (calc.)	VT10	17.2	0.16	0.44
b_{sp} (meas.)	VT2	4.3	3.22	0.81
b_{sp} (meas.)	VT.5	7.6	3.50	0.71

Values above are least squares best fit of form:

$y = y_0 + mx$ r = coefficient

210 measurements used for analysis
Parameters defined in text

Table 3

Fine Aerosol Species Concentrations,

1979, China Lake

Species	Average Concentration ug/m^3
Total mass	10.7 ± 4.8
Total carbon	1.71 ± 1.26
Soot	0.39 ± 0.30
$SO_4^=$	1.91 ± 1.21
NO_3^-	0.19 ± 0.18
NH_4^+	0.64 ± 0.44
Al	0.048 ± 0.069
Si	0.46 ± 0.24
S	0.76 ± 0.48
K	0.15 ± 0.10
Ca	0.086 ± 0.061
Fe	0.090 ± 0.049
Pb	0.076 ± 0.075
b_{sp}, $10^{-6}m^{-1}$	25.4 ± 13.2
b_{ap}, $10^{-6}m^{-1}$	5.1 ± 2.3

Average of 61 filter samples
Errors are standard deviations

To estimate the contribution of wind-blown dust of crustal origin to the fine aerosol concentration, elemental enrichment factors were calculated using the method of Macias et al. (20). The enrichment factor, EF_i, for an element i was calculated as follows:

$$EF_i = \frac{([M_i]/[Fe])_{aerosol}}{([M_i]/[Fe])_{crust}} \tag{5}$$

Iron was chosen as the reference element because its major source is likely to be soil and it is measured with good accuracy and precision by PIXE. Crustal abundances were taken from Mason (21). Enrichment factors greater than 1 indicate an enrichment of that element relative to crustal abundances; values less than 1 indicate a depletion. The results of this calculation are shown in Table 4. For this calculation it was assumed that ammonium and nitrate accounted for all aerosol nitrogen. It is seen that Si and Ca are near their crustal abundance, indicating a probable soil dust source. The low EF for Al is probably due to a systematic error in the Al measurement rather than a true depletion. Potassium, although present in small concentrations, is slightly enriched relative to crust. The other fine aerosol species, C, N, S, and Pb are enriched by factors of thousands over their natural crustal abundance, indicating that they are not due to wind-blown dust.

A chemical species mass balance for China Lake calculated using the 1979 average measured species concentrations is summarized in Table 5. The organic carbon concentration is taken to be 1.2 times the difference between the total and graphitic carbon concentrations. This was done to account for additional hydrogen and oxygen in the organic aerosol (22). The NO_3^- masses determined from the field samples were usually within one standard deviation of the average blank value. The low nitrate concentrations are consistent with measured gas phase nitric acid and ammonia concentrations measured from September 5-7, 1979. The concentrations of HNO_3 and NH_3 each averaged between 0.2 and 3 ppb at that time. At the temperatures measured, equilibrium considerations indicate no pure NH_4NO_3 should be present in the aerosol phase (23). The crustal aerosol concentration was computed from the Si, Ca, and Fe concentrations by assuming they were present in their bulk crustal abundance (21). It includes the contribution of species such as aluminum and oxygen, for which the data quality is poor, and is a reliable estimate of the contribution of wind-blown dust to the fine aerosol concentration.

It is seen that organics, sulfates, and crustal species contributed approximately equally to the measured fine mass concentration. However, the measured species accounted for an average of only 65% of the total measured mass concentration.

Table 4

Aerosol Elemental Enrichment Factors,

1979, China Lake

Element, X	$([X]/[Fe])_{aerosol}$	$([X]/[Fe])_{crust}$	Enrichment Factor, EF
C	19.1	4.0×10^{-3}	4.78×10^3
N	6.03	4.0×10^{-4}	1.51×10^4
Al	0.53	1.63	0.33
Si	5.12	5.54	0.92
S	8.51	5.2×10^{-3}	1.64×10^3
F	1.65	0.518	3.19
Ca	0.955	0.726	1.32
Fe	1.0	1.0	1.0
Pb	0.848	2.6×10^{-4}	3.26×10^3

Values computed using 1979 average species mass concentrations

Table 5

Fine Aerosol Species Mass Balance,

China Lake, California

Species	1979 Average Concentration $\mu g/m^3$
Organic	1.58
Soot	0.34
$SO_4^=$	1.91
NO_3^-	0.19
NH_4^+	0.64
Crustal	1.94
Other	0.3
Total Accounted	7.0
Total Measured	10.7

The most likely unmeasured species are Na and H_2O. If it were Na, however, it could not be in the form of NaCl or Na_2SO_4 since measured Cl concentrations were less than 0.1 $\mu g/m^3$ and SO_4^- was titrated with NH_4^+. The low average relative humidity, 35%, minimizes water condensation on the hygroscopic ammonium sulfate, but some water could have condensed on the filters during refrigeration. The filters were desiccated before and during weighing.

Elemental Size Distributions

Aerosol sulfur mass distributions were measured by the LPI-FVFPD technique on three separate occasions for 3 days each during 1978 and 1979. A total of 25 distributions were obtained. During each 3-day period the total sulfur concentration varied diurnally with the scattering coefficient, but the normalized sulfur mass distribution was preserved (13). However, the shape of the distribution was not preserved from period to period, as can be seen in Figures 5a-c. The dashed histogram in the figures is the inferred aerosol size distribution computed using the Twomey inversion algorithm to correct for the nonideal size resolution of the LPI (13). During the July 2-4, 1978 and September 4-7, 1979 periods, aerosol sulfur mass accumulated in particles between 0.05 and 0.5 μm aerodynamic diameter. The shapes of these distributions are quite similar to the total average volume distribution, Figure 4. The average b_{sp} during each of these periods was within one standard deviation of the annual average. Tracer studies have shown that the semi-arid Southern San Joaquin Valley was the source of the aerosulfate during the September 5-7, 1979 period (18).

During the October 20-22, 1978 period, the sulfur accumulated in larger particle sizes and in greater concentrations. This sampling period was characterized by high relative humidity (45-100%) associated with the aftermath of a tropical storm. It is likely that the source of the aerosol was the Los Angeles Basin where sulfate usually accumulates in particles between 0.5 and 1.0 μm diameter. The particle scattering coefficient during this episode was much higher than the average, often exceeding $100 \times 10^{-6} m^{-1}$, about 4 times the annual average. In examining the 1979 b_{sp}, temperature, and relative humidity data, it is clear that the July 1978 and September 1979 sample periods were more typical of the average than the October 1978 period. From these three periods, only the September 5-7, 1979 filter samples were included in the statistical analysis.

Elemental mass distributions were measured using the LPI-PIXE technique on September 7, 1979. A cyclone preseparator upstream of the LPI removed coarse aerosol. Some of the results are presented in Figures 6a-c. The concentration of

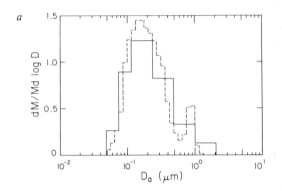

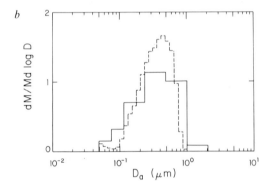

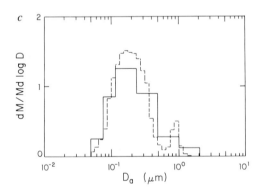

Figure 5. *Normalized S mass distribution, China Lake, CA—average of 8 samples. Aerosol segregated by LPI and analyzed by FVFPD. The solid histogram is the mass distribution with respect to the 50% aerodynamic cutoff diameter; the dashed histogram is the inverted distribution obtained from LPI calibration data and Twomey (1975) inversion algorithm: (a) July 2–4, 1978 (M = 0.765 μg/m³); (b) October 20–22, 1978 (M = 1.702 μg/m³); (c) September 5–7, 1979 (M = 0.954 μg/m³).*

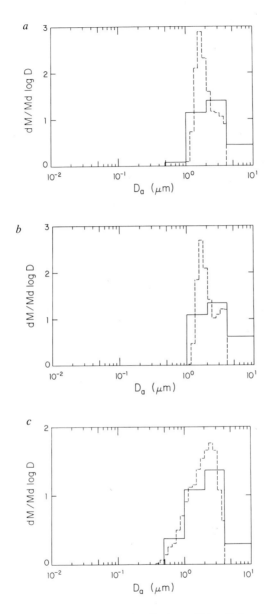

Figure 6. Normalized species mass distributions, China Lake, CA, 0823–1430 PST, September 7, 1979. Aerosol segregated by LPI and analyzed by PIXE. The solid histogram is the mass distribution with respect to 50% aerodynamic cutoff diameters; the dashed histogram is the inverted distribution obtained from LPI cali-bration data and Twomey (1975) inversion algorithm: (a) Si (M = 0.368 μg/m³); (b) Ca (M = 0.062 μg/m³); (c) Fe (M = 0.068 μg/m³).

coarse particles in these figures is reduced from the
ambient value because of the cyclone preseparator. Si, Ca,
and Fe accumulated in larger particle sizes, another indication
of the crustal origin of these elements.

The distribution of organic aerosol with respect to
particle size could not be measured with the LPI because of
the hydrocarbon anti-bounce coatings used.

Statistical Estimation of Species Mass Scattering Efficiencies

The goal of the year-long study at China Lake was to test
the applicability of statistical techniques to determine the
aerosol species contributions to the scattering coefficient.
On the average, the requirements for application of the statistical technique to filter data were met. Analysis of the 254
measured particle size distributions in 1979 indicates that
the fine aerosol volume distribution preserved its shape. The
measured sulfur mass distribution followed that of the total
submicron volume. By difference, it was assumed that the
organics did the same. The low relative humidity at China
Lake minimized the formation of aqueous solutions due to
water condensation on the particles. Therefore, it is expected
that the statistical technique can be used with some success
with the China Lake filter data.

As a first cut, univariate regression was used to see
relationships between pairs of measured variables. The results
are summarized in Table 6. It is seen that the particle
scattering coefficient is highly correlated with the total
fine aerosol mass concentrations, sulfate, and ammonium. To a
lesser degree it is correlated with the particle absorption
coefficient, potassium, and the unaccounted mass concentration,
M_u. The correlations between the particle scattering coefficient and the nitrate, total carbon (C_T) and crustal species
are poor. The poor correlation, $r=0.63$, between b_{ap} and the
soot concentration is probably an indication of the error in
the soot measurement. Although the sulfate and the unaccounted mass are highly correlated with the total mass, they are
not well correlated with each other ($r=0.49$).

On the basis of a preliminary regression analysis in which
it was found that the crustal species, e.g., Fe, Ca, and Si,
were well correlated with each other, the fine aerosol was
grouped into four species: sulfates, organics, crustal species,
and the unaccounted species. The concentrations of these four
species were calculated in the following manner:

1. $[sulfates] = [SO_4^=] + [NH_4^+]$

2. $[organics] = 1.2 \times [C_T]$

Table 6

Aerosol Mass Concentrations and b_{sp}: Results of

Univariate Regressions for China Lake, 1979

---

b_{sp} = 0.30 + 2.35 $[M_T]$ r = 0.86

b_{sp} = 7.5 + 9.36 $[SO_4^=]$ r = 0.86

b_{sp} = 8.9 + 25.8 $[NH_4^+]$ r = 0.85

b_{sp} = 4.2 + 4.13 b_{ap} r = 0.72

b_{sp} = 14.1 + 79.3 $[K]$ r = 0.58

b_{sp} = 14.3 + 2.41 $[M_u]$ r = 0.57

b_{sp} = 18.8 + 34.1 $[NO_3^-]$ r = 0.47

b_{sp} = 17.4 + 4.80 $[C_T]$ r = 0.46

b_{sp} = 15.6 + 114 $[Fe]$ r = 0.41

b_{sp} = 19.9 + 71.3 $[Ca]$ r = 0.32

b_{sp} = 18.2 + 17.1 $[Si]$ r = 0.31

b_{sp} = 3.2 + 4.5 $[Soot]$ r = 0.63

b_{sp} = 4.83 + 1.27 $[M_u]$ r = 0.81

b_{sp} = 4.3 + 3.33 $[SO_4^=]$ r = 0.83

b_{sp} = 2.20 + 1.26 $[SO_4^=]$ r = 0.49

Values computed from 61 filter samples

3. [crustal] = 21.6 x [Fe]

4. [unaccounted] = $[M_T]$ - [sulfates] - [organics]
    - [crustal]

Multiple regression analysis was applied to the four
species to seek a best statistical fit of the form:

$$\bar{b}_{sp} = \sum_{\alpha=1}^{N} \alpha_i M_i \qquad (6)$$

where M_i is the time average mass concentration of species i.
α_i could then be interpreted as the average mass scattering
efficiency of species i. The results of the technique applied
to the 61 China Lake filter samples are as follows:

b_{sp} = (0.97 ± 1.92) + (5.03 ± 0.64) [sulfates]

    + (1.54 ± 0.62) [organics] + (2.36 ± 0.67) [crustal]

    + (1.00 ± 0.31) [unaccounted]   (7)

    r=0.905

where b_{sp} is in units of $10^{-6}m^{-1}$ and the species concentrations
are in units of $\mu g/m^3$. The ± values in the above equation
represent one standard error in the coefficient estimates.
The contributions to the variance in b_{sp} were as follows:
sulfates,68.0%; organics, 6.8%; crustal, 13.9%; and unaccounted, 11.3%. The statistically inferred average mass scattering
efficiency for sulfates at China Lake in 1979 is compared to
values obtained by other investigators in Table 7. The
China Lake value is intermediate between those found in
other desert locations and in Los Angeles. The fine organic
aerosol was not statistically important in explaining the
variance in b_{sp}, and its inferred mass scattering efficiency
was less than one-third of the sulfate value. Similar results
were found by White and Roberts (24) in Los Angeles and
Trijonis and Yuan (3) in Phoenix. The crustal species in the
China Lake fine aerosol were, surprisingly, more important
than the organics in explaining the variance in the b_{sp},
but were less efficient per unit mass in scattering light.
     In summary, the variance in the measured fine particle
scattering coefficient, b_{sp} was dominated by sulfate concentrations. Organics and crustal species were much less
important statistically. The inferred mass scattering
effeciency for sulfates was intermediate between other desert
values and Los Angeles. The statistical results for China
Lake are quite similar to those by other investigators at
other locations, even though only the fine aerosol was sampled
at China Lake.

Table 7

Sulfate Average Mass Extinction Efficiency, α_i:

Comparison of Statistically Inferred Values

Source	Location	α_i, m^2/g
This study	China Lake	5.0
Trijonis & Yuan (3)	Phoenix	3.4
	Salt Lake City	4
White & Roberts (24)	Los Angeles	7
Cass (25)	Los Angeles	16
	Los Angeles	9 (nonlinear RH)
Waggoner et al (26)	Southern Sweden	5

Fine Aerosol Scattering Coefficient Balances: Comparison of
Statistical and Theoretical Results

The average contribution of each major aerosol species to
the 1979 average particle scattering coefficient, $\bar{b}_{spi}$ at
China Lake, was calculated using the following relationship:

$$\bar{b}_{spi} = \alpha_i \bar{M}_i \qquad (8)$$

where $\bar{M}_i$ is the 1979 average mass concentration of species i
and α_i is its average mass scattering efficiency. Both
statistical and theoretical techniques were used to calculate
α_i. The theoretical value of the average species mass
scattering efficiency was determined as the ratio of b_{spi}
calculated using Equation (1) to the measured species mass
concentrations. The measured aerosol composition size
distribution, $f_i(x)$, and calculated mass extinction effeciency,
$E_e(m_i, D, \lambda)$, were used in the determination of b_{spi}. The
sulfate mass scattering efficiency was computed using 25
mass distributions measured in July 1978, October 1978, and
September 1979. The average value was 3.20 m^2/g, slightly
greater than that of the total fine aerosol, 2.35m^2/g.
The mass scattering efficiency for the crustal species, i.e.,
Si, Ca, and Fe, in the fine aerosol, determined using
composition distributions obtained in September 1979, was
1.42 m^2/g. This value is much lower than that for sulfate
because the crustal species do not accumulate in the size
range which is optimum for scattering, i.e., 0.4 - 0.5 μm
diameter of a wavelength of 0.53 μm. In order to estimate
the mass scattering efficiency of the organic aerosols, it
was assumed that the organics accumulated in particles
smaller than 0.8 μm diameter and followed the 1979 average
measured volume distribution. Assuming a density of 1.7 g/cm^3,
the estimated organic mass scattering efficiency was 2.46 m^2/g,
intermediate between the values for sulfate and fine crustal
material. The unaccounted species have a mass scattering
efficiency of 2.35 m^2/g, assuming the species follow the
overall distribution. The uncertainties in each of these
values is estimated to be about 30%, primarily a result
of uncertainties in the measured mass distributions. The
results are shown in Table 8. It is seen that there is
excellent agreement between the average measured and calculated
b_{sp} using both techniques. Both techniques show that sulfate
is the most important identified light scattering species.
However, the statistical technique suggests a higher contribution of sulfate than does the theoretical model. The
contributions of organics and crustal species to b_{sp} are
each 10-20% for both techniques. The unaccounted species
contribute 39% to the theoretical scattering balance, approximately the same as its contribution to the fine mass concentration.

Table 8

Aerosol Species Contributions to the Particle Scattering Coefficient:

Comparison of Statistical and Theoretical Results for China Lake

Aerosol Species, i	1979 Average Mass Concentration, $\bar{M}_i$ μg/m³	Mass Scattering Efficiency, α_i m²/g		Contribution to b_{sp}, $10^{-6}m^{-1}$		% Contribution to Total Measured b_{sp}	
		Stat. (S.E.)	Theor. (S.E.)	Stat.	Theor.	Stat.	Theor.
Sulfates	2.55	5.03 (0.64)	3.20 (0.96)	12.8	8.2	50.4	32.3
Organics	1.97	1.54 (0.62)	2.46 (0.75)	3.0	4.8	11.8	18.9
Crustal	1.94	2.36 (0.67)	1.42 (0.42)	4.6	2.8	18.1	11.0
Unaccounted	4.24	1.00 (0.31)	2.35 (0.71)	4.2	9.9	16.5	39.0
Computed Total				24.6	25.7	97	101
Measured Total				25.4	25.4	100	100

Stat. - Statistical
Theor. - Theoretical
(S.E.) - Standard error of estimate

When the uncertainties in the values are included, it is found that the statistically determined mass scattering efficiencies are not significantly different than those calculated theoretically. For example, representing the uncertainty as twice the standard error results in a statistically inferred sulfate mass scattering efficiency of 5.0 ± 1.2 as compared with a theoretical value of 3.2 ± 1.9.

From Table 8 is is seen that roughly half the theoretically unaccounted light scattering is statistically associated with sulfate. The earlier fine mass balance indicated that not all the aerosol in the 0.05-0.5 μm diameter range could be accounted for by sulfates and organics. It is possible that the unaccounted species whose light scattering is statistically associated with sulfate may also be physically associated with sulfate in this size range. Water is the most obvious candidate, but is not likely to condense on $(NH_4)_2SO_4$ at the low relative humidities found at China Lake.

Sin summary, both techniques present a qualitatively similar picture of the contributions of aerosol chemical species to b_{sp}. Ammonium sulfate is the most important measured species, while the contributions of organics and crustal species are smaller, but not negligible. The species which could not be identified by chemical analysis also contributed to b_{sp}, but roughly half of this contribution is statistically linked to sulfate.

A Light Extinction Budget for China Lake

The light extinction coefficient, b_e, is the sum of the contributions of particles and gases to scattering and adsorption. Both fine and coarse particles contribute to light extinction. The nephelometer efficiently measures only the fine particle contribution to b_{sp}. From Figure 3 it is calculated that, on the average, coarse (D > 2 μm) particles contributed approximately 13 x $10^{-6}m^{-1}$ to b_{sp}. The median value will be somewhat less that this, since a small number of wind-blown dust episodes contributed the most to the average coarse particle contribution.

The calculated theoretical values were used to compute the 1979 China Lake light extinction budget shown in Table 9. As discussed earlier, the measured particle scattering coefficient varied diurnally, achieving maximum values at night and minimum values in early afternoon. For this reason Table 9 includes an estimated median 1 PM extinction budget. This may more closely reflect the people's perception of daytime visibility at China Lake. It is seen that, on the average, 51% of light extinction is due to Rayleigh scattering and wind-blown dust. Organics, soot, and sulfates contribute an average of 32% to extinction, while the remaining 17% is unaccounted. Interestingly, the carbonaceous aerosol

Table 9

1979 China Lake Light Extinction Coefficient Budget

Type of Light Extinction	Contribution to b_e, $10^{-6}m^{-1}$		Contribution to b_e, %	
	Average	Median 1 PM (est.)	Average	Median 1 PM (est.)
Rayleigh Scattering, b_{sh}	14	14	24.6	34.1
Absorption by gases, b_{ag} (est.)	0	0	0	0
Absorption by fine particles	5.1	3.6	8.9	8.8
Soot	5.1	3.6	8.9	8.9
Absorption by coarse particles	0	0	0	0
Scattering by coarse particles	13	5	22.8	12.2
Crustal species	13	5	22.8	12.2
Scattering by fine particles	25	18	43.9	43.9
Sulfate	8.2	5.9	14.4	14.4
Organics	4.8	3.4	8.4	8.3
Crustal species	2.8	2.0	4.9	4.9
Unaccounted	9.9	7.1	17.4	17.3
Total,	57	41	100	100
Average local visual range, km	68	95		

contributes more than sulfates to light extinction due to its ability to scatter and absorb light.

Using the calculated theoretical average mass extinction efficiencies, α_i, for each major chemical species, it is possible to estimate the percentage decrease in current visibility that would result from an increase in the mass concentration of each species. For a finite change in the mass concentration of a particular species, ΔM_i, the percentage decrease, $\Delta V_R/V_{RO}$ in current visual range, V_{RO}, is given by the following equation (13):

$$\frac{\Delta V_R}{V_{RO}} = \frac{V_{RO} \, \alpha_i \Delta M_i}{3.9 + V_{RO} \, \alpha_i \Delta M_i}$$

The calculations were performed assuming the median 1 PM visual range of 95 km and that 2 $\mu g/m^3$ of the species were added. The results are provided in Table 10. The reduction in visual range resulting from the increase in soil dust would be insignificant. The visibility reduction resulting from a 2 $\mu g/m^3$ increase in each of the other species would be significant, 11-33%. Soot, due to its efficient light absorption efficiency, would reduce the median 1 PM visual range at China Lake by one-third.

SUMMARY AND CONCLUSIONS

Multiple regression analysis was applied to the measured particle scattering coefficient and fine aerosol species mass concentrations from 61 filter samples collected at China Lake, California in 1979. Contributions of various aerosol species to the particle scattering coefficient were estimated. The statistically estimated contributions were compared with those determined theoretically using measured aerosol mass distributions.

It was found that the requirements were satisfied for application of the linear regression technique to species mass concentrations in a multicomponent aerosol. The results of 254 particle size distributions measured at China Lake in 1979 indicate that the normalized fine aerosol volume distribution remained approximately constant. The agreement between the calculated and measrued fine particle scattering coefficients was excellent. The measured aerosol sulfur mass distribution usually followed the total distribution for particles less than 1 µm. It was assumed that organic aerosol also followed the total submicron distribution.

The measured fine aerosol species were grouped into sulfates, organics, and crustal species, each having annual average mass concentrations of about 2-2.5 micrograms per cubic meter. An

Table 10

Estimated Reduction in Median 1 PMVisual Range at China Lake Due

to 2 μg/m³ Increase in Aerosol Species Mass Concentration

Aerosol Species	Average Mass Extinction Efficiency m^2/g	% Decrease in Current Visual Range Due to Additional 2 μg/m³ of Species
Soil dust (coarse and fine)	0.29	1.4
Ammonium Sulfate	3.2	13
Organics	2.5	11
Soot	10	33

Assumes current median 95 km visual range

average of 35% of the measured fine mass concentration could not
be accounted by chemical analysis.

The results of the statistically inferred species contributions to b_{sp} agreed qualitatively with those calculated
theoretically using measured aerosol distributions. Within
the errors in the measurements and analysis, the statistically
and theoretically determined values were not significantly
different. Ammonium sulfate was the most important fine
aerosol light scattering species, whereas the contributions
of the organics and crustal species were smaller, but not
negligible. Regression analysis estimated a somewhat higher
contribution of sulfate than that calculated theoretically.
Species which were not identified chemically also contributed
significantly to b_{sp}. The regression analysis indicated
that roughly half of the b_{sp} contribution by the unaccounted
species was statistically linked to sulfate. Water is sus-
pected but is probably not a significant component of the
aerosol at China Lake due to the low relative humidity. Using
measured 1979 values, a light extinction budget was calculated
for China Lake. On the average, 51% of light extinction at
0.53 μm wavelength was due to Rayleigh scattering and wind-
blown dust, in roughtly equal proportions. Fine carbonaceous
and ammonium sulfate aerosol contributed an average of 32%
to extinction, while the remaining 17% was unaccounted.

The quality of statistically inferred species extinction
balances can be enhanced with proper aerosol sampling. Due to
its important role in light scattering, only the fine aerosol
should be sampled. A mass balance should account for all
major fine particle species. Ideally the particle scattering
coefficient should be measured directly at the location where
aerosol is sampled by the filters. The importance of soot
and other carbonaceous aerosol contributions to light extinction in arid regions should not be overlooked.

Acknowledgements

This work was supported in part by National Science
Foundation grant number PFR76-04179, The Pasadena Lung
Association and the Naval Weapons Center at China Lake. We
thank Paul Owens and Tom Dodson of the Naval Weapons Center
for their assistance in collecting the experimental data.

Literature Cited

1. Middleton, W.E.K., Vision Through the Atmosphere, University
    of Toronto Press, Toronto: Canada, 1952.

2. Charlson, R.J.; Waggoner, A.P.; Thielke, J.F., "Visibility
    Protection for Class I Areas: The Technical Basis;" report
    to Council on Environmental Quality, August, 1978. Available
    from the National Technical Information Service, PB-288-842.

3. Trijonis, J.; Yuan, K., "Visibility in the Southwest: An Exploration of the Historical Data Base;" EPA-600/3-78-039,1978.

4. Ouimette, J.R.; Flagan, R.C., The Extinction Coefficient of Multicomponent Aerosols, accepted by Atmospheric Environment, 1980.

5. Wickramasinghe, N.C., Light Scattering Functions for Small Particles with Applications in Astronomy, John Wiley and Sons, New York, 1973.

6. Ouimette, J.R.; Friedlander, S.K.; Macias, E.S.; Stolzenkerg, M., "Aerosol Chemical Species Contribution to Visibility Reduction in the Southwestern United States;" Proceedings of the Grand Canyon Conference on Plumes and Visibility, November 10-14, 1980.

7. Ouimette, J.R., "Survey and Evaluation of the Environmental Impact of Naval Weapon Center Activities;" TM 2426, U.S. Naval Center. Available from National Technical Information Service, 1974.

8. Whitby, K.T.; Liu, B.Y.H.; Husar, R.B.; Barsic, N.J., The Minnesota Aerosol-Analyzing System Used in the Los Angeles Smog Project, J. Colloid and Interface Science, 1972, 39, 136.

9. Mathews, L.A.; Cronin, H.E., "Size Distributions of Atmospheric Aerosol at China Lake, California;" NWC Technical Memorandum TM 4109 (Unclassified), in preparation. Naval Weapons Center: China Lake, CA 93555, 1980.

10. Whitby, K.T.; Cantrell, B.K., "Electrical Aerosol Analyzer Constants," in Aerosol Measurement, D.A. Lundgren, S.S. Harris, Jr., W.H. Marlow, M. Lippmann, W.E. Clark, M.D. Durham, Eds, University Presses of Florida, Gainsville, Florida, 1979.

11. Friedlander, S.K., Smoke, Dust and Haze: Fundamentals of Aerosol Behavior, Wiley-Interscience, New York, 1977.

12. John, W.; Reischl, G., "A Cyclone for Size-Selective Sampling of Ambient Air;" J. Air Poll. Contr. Assoc. 30 872-876, 1980.

13. Ouimette, J.R., "Aerosol Chemical Species Contributions to the Extinction Coeffient;" Ph.D. Thesis, California Institute of Technology; Pasadena, California 1980.

14. Cahill, T.A., "Environmental Analysis of Environmental Samples," in New Uses of Ion Accelerators, J. Ziegler, Ed., 1-75, Plenum Press, NY, 1975.

15. Roberts, P.T.; Friedlander, S.K., Analysis of Sulfur in Deposited Aerosol Particles by Vaporization and Flame Photometric Detection, Atmospheric Environment, 1976, 10, 403.

16. Lin, C.; Baker, M.; Charlson, R.J., Absorption Coefficient of Atmospheric Aerosol; a Method for Measurement, J. Applied Optics, 1973, 12 (6), 1356.

17. Macias, E.S.; Radcliff, C.D.; Lewis, C.W.; Sawaki, C.R., Proton-Induced γ-ray Analysis of Atmospheric Aerosols for Carbon, Nitrogen, and Sulfur Composition, Anal. Chem., 1978, 50, 1120.

18. Reible, D.J.; Ouimette, J.R.; Shair, F.H., Visibility Degration Associated with Atmospheric Transport into the California Mojave Desert, accepted by Atmospheric Environment, 1981.

19. Sverdrup, G.M., "Parametric Measurement of Submicron Atmospheric Aerosol Size Distributions;" Ph.D. Thesis, Particle Technology Laboratory: University of Minnesota, Minneapolis 55455, 1977.

20. Macias, E.S., Blumenthal, D.L., Anderson, J.A., and Cantrell, B.K., Characterization of Visibility-Reducing Aerosols in the Southwestern United States: Interim Report of Project VISTTA, MRI 78-IR-1585, (1979).

21. Mason, B., Principles of Geochemistry, J. Wiley and Sons, Inc., New York, 1966.

22. Grosjean, D.; Friedlander, S.K., Gas-Particle Distribution Factros for Organic and Other Pollutants in the Los Angeles Atmosphere, J. Air Pollution Control Association, 1975, 25, 1038.

23. Stelson, A.W.; Friedlander, S.K.; Seinfeld, J.H., A Note on the Equilibrium Relationship between Ammonia and Nitric Acid and Particulate Ammonium Nitrate, Atmospheric Environment, 1979, 13, 369, (1979)

24. White, W.H. and Roberts, P.T., On the Nature and Origins of Visibility-Reducing Aerosols in the Los Angeles Air Basin. Atmospheric Environment, 11: 803, 1977.

25. Cass, G.R. On the Relationship between Sulfate Air Quality and Visibility with Examples in Los Angeles. Atmospheric Environment, 13: 1069, 1979.

26. Waggoner, A.P., Vanderpol, A.J., Charlson, R.J., Larsen, S., Granat, L., and Tragardh, C., Sulphate-Light Scattering Ratio as an Index of the Role of Sulphur in Tropospheric Optics. Nature, Lond., 261: 120, 1976.

RECEIVED May 11, 1981.

Aerosols from a Laboratory Pulverized Coal Combustor

D. D. TAYLOR and R. C. FLAGAN

California Institute of Technology, Pasadena, CA 91125

A laboratory study has been undertaken to characterize
the aerosol produced during pulverized coal combustion.
The emphasis in this work has been on the particulate
matter present in the flue gases at the inlet to the
gas cleaning equipment rather than that leaving the
stack. Coal is burned at conditions which simulate
the combustion region of coal-fired utility boilers.
The combustion products then pass through a series
of convective heat exchangers which cool them to
normal flue gas temperature. Samples extracted from
the cool (400-500°K) combustion products are analyzed
for major gaseous products and aerosol properties.
The size distribution of the particulate matter in
the 0.01-5 μm size range is analyzed on line using
an electrical mobility analyzer and an optical particle counter. Samples of particles having aerodynamic diameters between 0.05 and 4 μm are classified
according to size using the Caltech low pressure
cascade impactor. A number of analytical procedures
have been used to determine the composition distribution in these particles. A discrete mode of particles is observed between 0.03 and 0.1 μm. The major
components of these particles are volatile elements
and soot. The composition of the fine particles
varies substantially with combustor operating conditions.

Fine particles from coal combustion are a threat to human
health and air quality. Numerous studies have reported higher
concentrations of volatile trace elements in small particles

than in larger particles (1-6). This has been attributed to
the condensation or adsorption of vapors on particle surfaces.
The fine particles are also enriched with condensed hydrocarbons and have higher mutagenic activity than do larger particles (7).

Most of the published composition/size distribution data
have been obtained by analyzing cascade impactor samples. Some
of these data suffer from poor size classification as a result
of particle bounce or reentrainment, seriously limiting size
resolution. Even when this problem is overcome, the data obtained with conventional cascade impactors are not capable of
resolving many details of the distribution of submicron particles. These instruments typically classify only those particles larger than 0.3-0.5 µm aerodynamic diameter. All smaller
particles are collected on a filter downstream of the impactor.
Some measurements of the variation of composition with size
below this limit have been attempted by aerodynamically classifying resuspended ash (8). These data suffer from incomplete
disaggregation as well as poor classification of the smaller
particles.

Particle size distributions of smaller particles have been
made using electrical mobility analyzers and diffusion batteries, (9-11) instruments which are not suited to chemical characterization of the aerosol. Nonetheless, these data have made
major contributions to our understanding of particle formation
mechanisms (12,13). At least two distinct mechanisms make major contributions to the aerosols produced by pulverized coal
combustors. The vast majority of the aerosol mass consists of
the ash residue which is left after the coal is burned. At
the high temperatures in these furnaces, the ash melts and co-
alesces to form large spherical particles. Their mean diameter is typically in the range 10-20 µm. The smallest particles
produced by this process are expected to be the size of the
mineral inclusions in the parent coal. Thus, we expect few
residual ash particles smaller than a few tenths of a micro-
meter in diameter (12).

The observed number distribution of the aerosol leaving a
pulverized coal combustor peaks at about 0.1 µm diameter (9-11).
Careful analysis of these data reveals that the fine particles
form a narrow mode in the mass distribution, accounting for a
few percent or less of the total aerosol mass. Because gas
cleaning devices such as electrostatic precipitators have a
relatively low collection efficiency in the 0.1-1 µm size range,
these fine particles pass through the collection equipment much
more readily than do larger particles, and the relative amount
in the stack effluent is greatly enhanced. In one study, this
fine particle mode accounted for 19 percent of the stack aerosol even though they only amounted to 2 percent of the mass of
aerosol leaving the boiler (11). Any chemical species which

are concentrated in these particles will contribute a dispro-
portionate amount of the aerosol emitted to the atmosphere.
The fine ash particles are generally thought to result
from nucleation of condensing ash vapors (12). Because of the
very high temperatures in pulverized coal flames (T > 2000 K),
even refractory constitutents of the ash such as Si and Al may
be volatilized (14-16). The new particles formed by vapor condensation are initially very small, but grow by coagulation and
condensation to produce the narrow peak in the mass distribution
in the 0.01-0.1 μm size range. The composition of the fine new
particles may differ substantially from that of the bulk ash.
Volatile ash species may be concentrated in nucleation generated
particles to a much greater extent than would be expected if the
only enrichment mechanism were heterogeneous condensation. There
has been some speculation that most of the fine particles are
produced by the physical disruption of large ash particles, particularly by the bursting ash bubbles known as cenospheres (8). The
composition of the small particles formed by this mechanism is
expected to be similar to that of the bulk ash, but enriched by
heterogeneous condensation of volatile species. Although some
atomization of molten ash by this mechanism has been observed (17),
there is little evidence that bursting of bubbles will produce
large numbers of particles smaller than a few tenths of a micron
in diameter (18).
    The composition distribution of the particles produced in a
laboratory pulverized coal combustor will be explored in this
paper using aerosol classification techniques capable of resolving the composition distribution to 0.03 μm diameter. Unlike
previous attempts to measure the composition distribution, the
particles were classfied directly, without having to resort to
resuspension, using calibrated instruments. Experiments were
conducted in a laboratory combustor in which operating parameters
can be varied over a wide range. Data are presented which demonstrate that the composition of fine particles varies substantially
with combustion conditions and does, under some conditions, differ
considerably from that of the bulk ash.

Experimental Laboratory Coal Combustor

    The laboratory furnace, illustrated in Figure 1, has been
described in detail elsewhere (19). The combustion chamber
design is similar to that of Pershing and Wendt (20). It consists of a vertical cylinder 1.0 m long and 0.2 m inside diameter cast from alumina refractory cement. A series of convective heat exchangers, also 1.0 m long and 0.2 m inside
diameter, are mounted directly below the combustor. The combustor is fired at a rate of 8 to 12 Kw, providing a residence
time of 1 to 2 seconds in the combustion chamber.

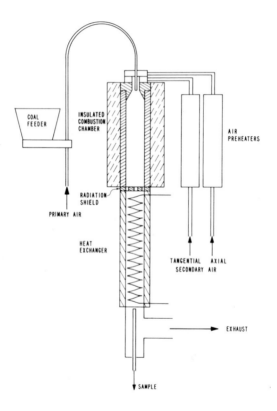

Figure 1.   Schematic of the laboratory pulverized coal combustor

Pulverized coal is fed directly from a variable speed auger into the high velocity primary air stream which conveys it to the injector at the top of the furnace. The coal and primary air enter the combustor through a single low-velocity axial jet. Secondary combustion air is divided into two flows which enter the combustor coaxial to the primary stream. Part of the flow is introduced through a number of tangential ports to induce swirl which is necessary for flame stabilization. The remainder enters the combustor axially. The two secondary air streams are separately preheated using electrical resistance heaters. The swirl number, a dimensionless ratio of the angular momentum to the product of the axial momentum and the radium of the burner, can be varied through separate control of the two secondary air streams in order to study various burner designs. The air flows were measured using sharp edged orifices. Control of the air flows and calibration of the coal feeder made it possible to duplicate combustion conditions as determined both by exhaust gas analysis (CO, CO_2, O_2, NO, NO_x) and aerosol characteristics. The fuel burned in the present experiments was a Utah subbituminous coal containing 10.16 percent ash. Its properties are summarized in Table I.

Aerosol Sampling and Characterization. Aerosol samples were extracted downstream of the convective heat exchangers using uncooled probes. The probe was designed to minimize collection of particles larger than 20 μm and to quench the coagulation of the fine particles, see Figure 2. Aerosol enters the probe through slots in the side of the probe. Excess sample is extracted from the flow in order to achieve the desired aerodynamics. The flow is then split. A small fraction passes through a conical tipped capillary tube. The pressure drop across this laminar flow element is monitored as a rough measure of the sample flow. At the capillary outlet, the flow is diluted with clean, filtered air to quench coagulation and to minimize condensation. The dilution ratio is determined by measuring the NO_x concentration in both the undiluted and diluted gas streams. Uncertainty in these determinations is estimated to be 20 percent. All data are corrected for dilution.

On-line aerosol measurements were made using a Thermo-Systems, Inc., Model 3030 Electrical Aerosol Size Analyzer (EAA). This instrument uses the electrical mobility of the particles to measure the size distribution in the 0.01 to 0.5 μm range.

Aerosol samples for composition measurements were collected using the Caltech low pressure impactor (LPI) (21,22). This fully calibrated instrument classified particles into eight size fractions between 0.05 and 4 μm aerodynamic diameter. Collection of particles smaller than 0.3 μm is achieved by reducing the pressure and increasing the velocity of the jets.

Table I

Analysis of Coal and Ash Samples

Ultimate Analysis - Utah II Bituminous Supplied by EER, Inc.

Substances	Dry Basis, wt fraction
Carbon	.72
Hydrogen	.058
Oxygen	.102
Sulfur	.0076
Ash	.095

Ash Analysis - X-ray Fluorescence

Na_2O	.0061
K_2O	.0094
CaO	.0492
MgO	.0087
Al_2O_3	.1314
SiO_2	.5900
Fe_2O_3	.0358
TiO_2	.0067
SO_3	.1646

Higher Heating Value of Coal - 2.98×10^7 J/kg

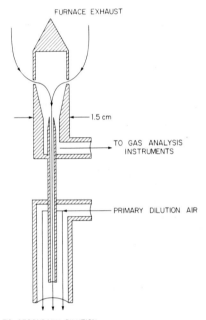

FURNACE EXHAUST

1.5 cm

TO GAS ANALYSIS INSTRUMENTS

PRIMARY DILUTION AIR

TO SECONDARY DILUTION
AND AEROSOL INSTRUMENTS

Figure 2.   Schematic of the dilution sampling probe

The stages which collect the two smallest particle sizes have
sonic jets.
The aerosol samples are impacted on greased substrates to
minimize particle bounce. Particles may bounce even from well
greased surfaces if they impact at too high a velocity. This
problem is reduced by taking care that particles much larger
than the critical impaction diameter are not encountered at
any stage of the impactor. The LPI size cuts are closely
spaced to reduce particle bounce, but for the aerosol being
studied here additional precautions must be taken. The maximum cut size for the LPI is 4 μm. Because the mass mean diameter of the ash aerosol is 10 to 20 μm, the vast majority of
the aerosol mass must be removed from the sample before it
enters the impactor. Otherwise, the samples of small particles
would be contaminated with reentrained large particles. Two
cyclone separators were used in series to remove particles
larger than 2.2 μm. These instruments were built to the specifications of the California Air Industrial Hygiene Laboratory
respirable particle sampler (23).
Particle bounce or reentrainment also becomes a problem
if the impactor stages are overloaded. This complicated the
collection of reasonable quantities of fine particles since
care must be taken to avoid overloading the upper impactor
stages with large particles. In some of the experiments reported here, duplicate stages for collection of 4, 2, and 1 μm
particles were assembled in series to insure that large particles would not bounce through the impactor and contaminate the
samples of fine particles. The quantity of material collected
on the duplicate stage was small for each of these size cuts.
The elemental analysis of the LPI samples was performed
by particle induced X-ray emission analysis (PIXE) at the
Crocker Nuclear Laboratory of the University of California at
Davis. The methods for sample preparation and analysis were
those developed by Ouimette (24). The LPI samples were collected at the center of a 13 mm disk of 3.2 μm thick mylar which
was fastened to 25 mm glass collection plates. The mylar was
cleaned with spectrograde toluene and decharged using a radioactive source. It was then greased by depositing 0.5 μℓ of a
solution containing 2 percent by weight Apiezon L in toluene
on the center of the disk. After samples were collected, the
mylar disks were transferred to 35mm slide frames, mounted in
slide trays, and shipped to UC Davis for analysis. Blank samples were prepared in the same manner and included among the
samples. The precision of the measurements on repeated analyses of prepared standards was typically $\pm$ 8 to $\pm$ 17 percent.
Semiquantitive estimates of the quantity of graphitic carbon
in the particles was made by optical absorption measurements
using a modified integrating plate technique developed by
Ouimette (24).

Results

Aerosol measurements have been made burning the same coal in two combustor operating modes. Early in our experimental program, a conical coal injector tip was used. High swirl numbers (S = 4.3) were required to stabilize the flame and achieve acceptable combustion efficiencies. This resulted in the centrifugal deposition of some of the coal on the combustor walls, so only a fraction of the coal was burned in suspension. This influences the quantity of coal burned at high temperature and, thus, the quantity of ash vaporized. In these early experiments a radiation shield was also installed between the combustion chamber and the heat exchangers to reduce the radiative heat loss from the flame region. As a result, the maximum measured wall temperature was high, i.e., 1260°C at a fuel-air equivalence ratio, ϕ = 0.96.

The aerosol composition distribution, Figure 3, shows pronounced variation with particle size. The distribution has a distinct break in the 0.1 to 0.3 μm size range. The larger particles account for the vast majority of the aerosol mass. Only a small fraction of these particles is included in the impactor sample because the cyclone separator removed most particles larger than about 2 μm. The composition of those large particles which were collected by the impactor is that expected for the ash residue. The major species are the oxides of Si, Al, Fe, and Ca. Sulfur, sodium, and other minor species only accounted for a small fraction of these particles. Some carbon was also found in these large particles. This is not surprising: imperfect mixing of fuel and air makes it unlikely that all of the char would be consumed at a fuel-air ratio so near stoichiometric.

The fine particle mode also contains oxides of Si and Fe, along with smaller amounts of Ca and Al. In contrast to the larger particles, sulfur (presumably in the form of sulfates), carbon (or soot), and sodium (possibly Na_2O or Na_2SO_4) are major species in the fine particles. The estimates of carbon in these samples are highly uncertain because the absorbtivity was 50 to 70 percent. Although sodium was clearly present in these samples, the standard deviations were large, from 40 to 260 percent. The sulfur, sodium, and carbon were almost certainly deposited in the fine particles by vapor condensation. The more refractory species, Si, Fe, Ca, and Al, may also have accumulated in the fine particles as a result of vaporization in the high temperature flame region and later condensation. The sharp break between the two modes, in terms of both mass and composition, is strong evidence that the particles in the two modes are formed by different mechanisms.

The mass distribution inferred from the EAA data is also shown in Figure 3. The two determinations are in reasonable

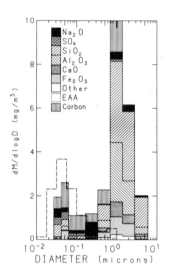

Figure 3. Composition distribution of the aerosol produced by near stoichiometric combustion (φ = 0.96) and swirl number of 4.3. Oxides were determined as elements using PIXE; carbon was inferred from absorption measurements.

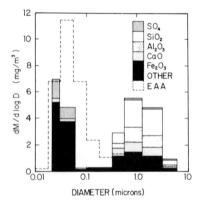

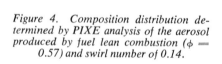

Figure 4. Composition distribution determined by PIXE analysis of the aerosol produced by fuel lean combustion (φ = 0.57) and swirl number of 0.14.

qualitative agreement given the uncertainties in the PIXE analysis, carbon measurement, sample dilution ratios, and the possibility that species which are not measured, such as water, may be important.

The conical coal injector was replaced with a blunt cyclinder with a single axial jet so the flame could be stabilized at lower swirl numbers, thereby reducing the centrifugal deposition on the furnace walls. The radiation shield between the combustor and heat exchanger was removed to reduce particle losses further. The increased radiative transfer decreased the wall temperature substantially. The later experiments were also carried out at lower fuel-air equivalence ratios, i.e., ϕ = 0.57. The combination of increased heat losses and increased dilution with excess air reduced the maximum wall temperature to 990°C for the experiments reported below.

The combination of lower temperatures and higher oxygen concentrations resulted in a substantially different composition distribution, as can be seen in Figure 4. Particles larger than 0.1 μm are similar in composition to those described above quantity of char does, however, appear to be much smaller. The fine particles from fuel lean combustion contained no measureable soot. Whereas the submicron particle samples from the previous experiment varied from black to gray, the present samples ranged from clear to red. The smaller quantities of soot and char can be attributed directly to the presence of sufficient oxygen to complete the combustion process in spite of imperfect mixing of fuel and air. The differences in the compositions of particles in the two modes are much more apparent than was seen in Figure 3. Here the fines are composed predominantly of iron and contain no measureable silicon or aluminum.

The concentrations (mass of element/total mass of oxides) of a number of elements are plotted as a function of particle size in Figure 5. Analyses of two sets of impactor samples collected at the same combustion conditions on different days are reported. The circles and left facing error bars are data from a fairly heavily loaded impactor, 40 μg total sample. The squares and right facing error bars are from a more lightly loaded sample, 18 μg total sample. The two measurements are in close agreement, within one standard deviation for most of the eight samples of six elements collected on each of two impactors, suggesting that the measurements are, at the very least, consistent from one sample to another. The large uncertainty in the intermediate size range (0.08 to 0.4 μm) is due to the very small quantities of material on these impactor stages (< 1 μg).

The most refractory species, Al and Si, show little variation with size for particles larger than 0.08 μm. The concentrations of both species are very similar to the quantities determined by X-ray fluorescence analysis of bulk ash samples, indicated by the broken lines. The smaller particles contain very little of these major constitutents of the ash, most of

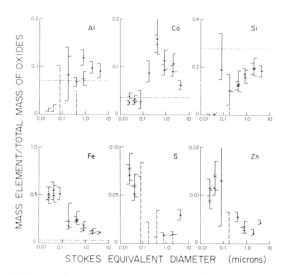

Figure 5. Variation of element concentrations with particle size for $\phi = 0.57$, $S = 0.14$. Circles and left-facing error bars are from a heavily loaded LPI. Squares and right-facing error bars are from a lightly loaded impactor. Error bars indicate standard errors or lower detection limits. Broken lines are the element concentration in the bulk ash.

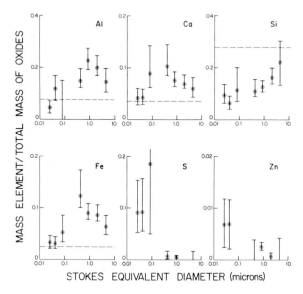

Figure 6. Variation of element concentrations with particle size for $\phi = 0.96$, $S = 4.3$. Error bars indicate standard errors or lower detection limits. Broken lines are the element concentrations in the bulk ash.

the data are below the detection limit of PIXE analysis. The very sharp decrease of more than an order of magnitude in these concentrations is a clear indication that the mechanism by which the very small particles are formed is distinctly different from the residual ash formation mechanism. In addition, the low concentrations of the two major constitutents of the bulk ash is a clear indication that the fines are not produced by mechanical break-up of the larger particles.

Iron, zinc, and sulfur concentrations increase with decreasing particle size, as suggested by the vapor condensation mechanism. Even particles a few microns in diameter contain much more iron than does the bulk ash, suggesting that they are enriched by heterogeneous condensation. The concentration trend extends at least to 0.05 μm. Below this size the concentration trend appears to level off, as would be expected if the smaller particles had undergone substantial coagulation after being formed by nucleation. The data for zinc show a trend similar to that of iron. The high concentration on the first impactor stage (4 μm) may be the result of unburned char carryover, some large (20 μm) char particles were observed in this sample. The increase in the concentration of both iron and zinc amounts to about a factor of 5 over the size range of the LPI. The variation of the sulfur concentration is similar to that of zinc.

Calcium exhibits a concentration variation intermediate between these two groups of elements. Its concentration increases slightly with decreasing particle size for particles larger than 0.1 μm, and then drops by about half an order of magnitude. This suggests that some calcium is vaporized, but the amount is small compared to iron.

Similar plots of the data from higher temperature, nearly stoichiometric combustion, Figure 6, show substantially different trends. The refractory species, i.e., Al, Si, and Ca, show little variation with particle size. Comparison with Figure 5 suggests that this results from vaporization of small amounts of these species. The apparent decrease in the fine particle iron is caused by dilution with other major ash constitutents. Both sulfur and zinc concentrations increase as size decreases in spite of the larger amount of the other species in the fines.

Discussion and Conclusions

The aerosol produced by a laboratory pulverized coal combustor was size classified in the range 0.03 to 4 μm Stokes equivalent diameter using a low-pressure cascade impactor. The samples thus collected were analyzed using a focussed beam particle induced X-ray emission technique. This combination of techniques was shown to be capable of resolving much of the structure of the submicron coal ash aerosol. Two distinct modes in the mass distribution were observed. The break between these modes was at a particle size of about 0.1 μm.

The composition of the particles larger than 0.1 μm was similar to that of the bulk ash. The fine particle composition (D < 0.1 μm) varied. Fuel lean combustion at relatively low combustor wall temperatures produced fines which contained no measureable Al or Si. The lack of the two major components of the coal ash supports the hypothesis that the majority of these small particles were not formed by the mechanical break-up of larger ash particles, but rather by nucleation of condensing ash vapors. The major component of the fine particles was iron. The smaller particles were also significantly enriched with zinc and sulfur, the most volatile species for which composition distributions were obtained in the present study. In contrast to Davison, et al., (1) the concentration of calcium was also found to increase somewhat with decreasing particle size for particles larger than 0.1 μm. Its concentration decreased below this size.

High temperature combustion at a nearly stoichiometric fuel-air ratio produced substantially different trends. While the behavior of sulfur and zinc did not change significantly, the fine particles contained large amounts of aluminum, silicon, and calcium. Iron made up a relatively small amount of the fine particles in this case. The difference between the two experiments is probably due to increased vaporization of the refractory species at higher combustion temperatures. The relative decrease in fine particle iron may result from dilution with silicon and, to a lesser extent, aluminum and calcium.

Although the data presented here are limited to a single coal burned in two combustor operating modes, several important observations can be made about the fine particles generated by pulverized coal combustion. The major constituents of the very small nucleation generated particles vary with combustion conditions. High flame temperatures lead to the volatilization of refractory ash species such as silica and alumina, probably by means of reactions which produce volatile reduced species such as SiO or Al. At lower flame temperatures which minimize these reactions other ash species dominate the fine particles. Because the major constitutents of the fine particles are relatively refractory, nucleation is expected to occur early in the combustion process. More volatile species which condense at lower temperatures may also form new particles or may condense on the surfaces of the existing particles. Both mechanisms will lead to substantial enrichment of the very small particles with the volatile species, as was observed for zinc.

Because the aerosol size and composition distributions depend so strongly on the combustion conditions, substantial differences in fine particle formation and emissions are expected between different furnaces, fuels, and operating conditions with the greatest variation in particles smaller than a few tenths of a micron in diameter.

Acknowledgements

The authors thank Dr. James Ouimette for assistance in collecting aerosol samples for analysis. This work was supported by National Science Foundation Grant PFR 76-04179 and the Pasadena Lung Association.

Literature Cited

1. Davison, R.L.; Natusch, D.F.S.; Wallace, J.R.; Evans, Jr., C.A., Environ. Sci. Techn. 1974, 8, 1107-1113.

2. Kaakinen, J.W.; Jorden, R.M.; Lawajani, M.H.; West, R.E., Environ. Sci. Techn.1975, 9, 862-869.

3. Lee, R.E., Jr.; Christ, H.L.; Riley, A.E.; MacLeod, K.E., Environ. Sci. Techn. 1975, 9, 643-647.

4. Block, C.; Dams, R., Environ. Sci. Techn. 1976, 10, 1011-1017.

5. Gladney, E.S.; Small, J.A.; Gordon, G.E.; Zohler, W.H., Atmos. Environ. 1976, 10, 1071-1077.

6. Ragaini, R.C.; Ondov, J.M.; Radio, J., Anal. Chem. 1977, 37, 679-691.

7. Fisher, G.L.; Chrisp, G.E.; Raabe, O.G., Science 1979, 204, 879-881.

8. Smith, R.D.; Campbell, J.A.; Nielson, K.K., Environ. Sci. Techn. 1979, 13, 553-558.

9. McCain, J.D.; Gooch, J.P.; Smith, W.B., J. Air Poll. Contr. Assoc. 1975, 25, 117-121.

10. Schmidt, E.W.; Gieseke, J.A.; Allen, J.M., Atmos. Environ. 1976, 10, 1065-1069.

11. Ensor, D.S.; Cowen, S.; Hooper, R.; Markowski, G., Evaluation of the George Neal No. 6 Electrostatic Precipitator, Electric Power Research Institute Report No. EPRI FP-1145, 1979.

12. Flagan, R.C.; Friedlander, S.K.; Shaw, D.T., Ed. "Recent Developments in Aerosol Science"; Wiley, 1978, pp. 25-59.

13. Flagan, R.C. Seventeenth Symposium (International) on Combustion, The Combustion Institute, 1979, pp. 97-104.

14. Sarofim, A.F.; Howard, J.B.; Padia, A.S., Combust. Sci. Techn. 1977, 16, 187-204.

15. Desrosiers, R.E.; Rieehl, J.W.; Ulrich, G.D.; Chiu, A.S. Seventeenth Symposium (International) on Combustion, The Combustion Institute, 1979, pp. 1395-1403.

16. Mims, C.A.; Neville, M.; Quann, R.J.; Sarofim, A.F. Laboratory Studies of Trace Element Transformations During Coal Combustion, presented at the 87th National AIChE Meeting, Boston, Mass., August 1979.

17. Raask, E., J. Inst. Fuel 1969, 44, 339-344.

18. Tomaides, M.; Whitby, K.T.; Liu, B.Y.H., Ed. Bursting of Single Bubbles, "Fine Particles"; Academic Press: New York, 1976; pp. 235-252.

19. Flagan, R.C.; Taylor, D.D. Laboratory Studies of Submicron Particles From Coal Combustion, Eighteenth Symposium (International) on Combustion, Waterloo, Ontario, Canada, August 1980 (in press).

20. Pershing, D.W.; Wendt, J.O.L. Sixteenth Symposium (International) on Combustion, The Combustion Institute, 1977, p. 389.

21. Hering, S.V.; Flagan, R.C.; Friedlander, S.K. Environ. Sci. Techn. 1978, 12, 667.

22. Hering, S.V.; Friedlander, S.K.; Collins, J.; Richards, J.W. Environ. Sci. Techn. 1979, 13, 184.

23. John, W.; Reischl, G.A. Cyclone for Size Selective Sampling of Ambient Air, California Department of Health Air Industrial Hygiene Laboratory Report, No. 187, June 1978.

24. Ouimette, J.R. Aerosol Chemical Species Contribution to the Extinction Coefficient, Ph.D. Thesis, California Institute of Technology, July 1980.

RECEIVED April 24, 1981.

Elemental Composition of Atmospheric Fine Particles Emitted from Coal Burned in a Modern Electric Power Plant Equipped with a Flue-Gas Desulfurization System

J. M. ONDOV, A. H. BIERMANN, R. E. HEFT, and R. F. KOSZYKOWSKI

Environmental Sciences Division, Lawrence Livermore Laboratory,
P.O. Box 5507, L-453, Livermore, CA 94550

Improved control devices now frequently
installed on conventional coal-utility boilers
drastically affect the quantity, chemical
composition, and physical characteristics of
fine-particles emitted to the atmosphere from these
sources. We recently sampled fly-ash aerosols
upstream and downstream from a modern lime-slurry,
spray-tower system installed on a 430-Mw(e) coal
utility boiler. Particulate samples were collected
in situ on membrane filters and in University of
Washington MKIII and MKV cascade impactors. The MKV
impactor, operated at reduced pressure and with a
cyclone preseparator, provided 13 discrete
particle-size fractions with median diameters
ranging from 0.07 to 20 μm; with up to 6 of the
fractions in the highly respirable submicron
particle range. The concentrations of up to 35
elements and estimates of the size distributions of
particles in each of the fly-ash fractions were
determined by instrumental neutron activation
analysis and by electron microscopy, respectively.
Mechanisms of fine-particle formation and chemical
enrichment in the flue-gas desulfurization system
are discussed.

It is now well documented that large numbers of particles
are emitted from coal-fired power plants in distinct
distributions with modal diameters less than 1 μm (1-5).
While major components of those aerosols are thought to be
oxides of Si, Al, Na, and Ca (5-6), the concentrations of most
of the other elements have not been accurately determined. As
predicted by models of gas-to-particle deposition (2,7-9),
experimental data (8-11) indicate that the concentrations of
vapor-deposited species in these very fine particles are often

0097-6156/81/0167-0173$05.00/0

10- to 100-fold greater than those of the large-particle region of the distribution. Theoretically, at least some of the particles may be homogeneously condensed and contain a single chemical component.

Although the release of large numbers of these highly chemically enriched particles may have profound health and environmental consequences, the published literature contains virtually no data characterizing the distribution, physical properties, chemical composition, and source terms of a broad range of important primary pollutants emitted in submicrometer particles.

In previous studies we found that improved emission-control devices (such as hot-side electrostatic precipitaters and wet-scrubber systems) now being installed on modern pulverized-coal-fired power plants modify the quantity, chemical composition, and distribution characteristics of fine aerosol emissions (12,13). Such modifications must be understood to adequately assess human health and environmental hazards, and to apportion the contributions of sources to urban pollutant inventories.

In this work, we use a University of Washington low pressure impactor (LPI) and instrumental neutron activation analysis (INAA) to determine the elemental composition of aerosols from a two 430 MWe coal-utility boilers, ranging in diameter from less than 0.07 to about 10 μm, and to investigate the modification of the aerosol by a modern flue-gas desulfurization system. A preliminary account of the work is presented here.

Experimental Measurements

Plant description. Two nearly identical 430-Mw(e), western, conventional pulverized-coal-utility boilers (referred to as plants A and D) were tested. Both units use tangentially fired burners and burn low-sulfur 200-mesh coal of heat content approximately 27 000 J/g. Both units are equipped with cold-side electrostatic precipitators (ESP) of design efficiency of 99.5% or greater, and a modern flue-gas desulfurization (FGD) system consisting of four verticle spray towers.

Each adsorber tower contains a series of spray nozzles and a mist eliminator as shown in Figure 1. After contacting the flue gas, the scrubbing solution (a 6% by weight calcium oxide-water slurry) collects in the bottom of the tower where it is continuously stirred. Fresh lime slurry is added at the bottom of the tower where it mixes with the liquid injected into the spray section. This mixed liquid is continuously recycled to the spray section, and functions as the adsorbing agent. Overall reduction in the SO_2 concentration is designed to be 80% when 10% of the flue gas is bypassed. The system is designed to limit the emission of total suspended particles

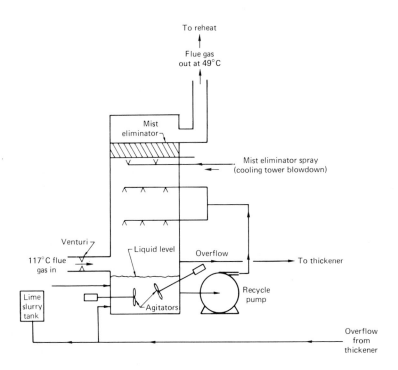

Figure 1. Flow schematic of a spray-tower adsorber. The scrubbing solution is contacted with hot flue gas, collected in the bottom, and continuously recycled and contacted. Suspended solids and pH of liquid in the recycle loop of Plant D spray towers ranged from 5.2 to 8.7%, and from 5.2 to 6.80%, respectively.

(TSP) to 0.032 g/m³, provided the inlet TSP concentration does
not exceed that level.

Sampling. Four aerosol samples were collected
isokinetically on July 26, 1979, at ports on the inlet duct of
the ESP at Plant A; two were taken with 62-mm fluoropore filters
and two were taken with the University of Washington MKV Cascade
impactor (14). At plant D, samples were collected over a 6-day
period at ports both in the outlet duct of the ESP (i.e.,
upstream of the FGD system) and at the 91-m level of the stack.
Eleven fluoropore filter, 1 MKV, and 4 MKIII impactor samples at
each location, giving a total of 22 filter, and 8 MKIII
samples. A single MKV sample was also collected in-stack at
reduced pressure at plant D during the 6-day period.
Polycarbonate material coated with apiezon L vacuum grease and
62 or 47-mm-diam, 1-μm pore Fluoropore filters were used as
back-up filters in the MKV impactor.

The pH and solids content of liquid in the recycle loop of
each spray tower and of the lime slurry were monitored and
recorded by plant personnel at 4-h intervals during the entire
test period. The pH and solids content of the recycled liquid
ranged from 5.2 to 6.8 and 5.2 to 8.7%, respectively; for the
lime slurry, the ranges were 4.8 to 5.9, and 12.2 to 12.8%,
respectively.

Operation of the University of Washington MKV Impactor.
The MKV is a multicircular jet impactor similar in design to the
MKIII, but it has 11 impactor stages and may be operated as a
high-pressure drop impactor (15), with an out-board back-up
filter holder. The orifice plates of the last four stages of
the MKV are quite similar to those used in the University of
Washington MKIV, which was designed specifically for low-
pressure operation. With a precyclone, the unit can provide 13
discrete particle size fractions, with up to 6 of the fractions
in the highly respirable submicrometer particle-diameter-range.
Unlike the MKIV, the MKV can be inserted into the duct through
standard 4-in i.d. sampling ports, so that the aerosol is sized
in situ without dilution or heating. A rotatable joint
(described in Reference 11) placed between the probe and the
sampler allows us to turn the sampler into the gas flow so that
a straight, rather than curved ("gooseneck"), nozzle may be
used. In low-pressure operation, the absolute pressure of gas
in the last stage is monitored throughout the run.

Theoretical efficiency curves for each stage may be
calculated if the absolute pressure of the gas in each of the
stages is known. Pilat et al. (16) have extensively measured
the pressures on each stage of their MKIV impactor as a function
of gas flow, and have constructed an empirical relationship for
calculating the pressures of preceeding stages, given the
pressure of the final stage and the overall flow rate of gas
through the impactor. Pilat and Mummey (15) used this empirical

relationship to prepare a set of theoretical calibration curves specifically for the final stage pressures and flow rates of our MKV unit. Because of the constraints of isokinetic sampling, and the somewhat high-pressure-drop-filter required for our chemical analyses, we were unable to duplicate the final stage pressures and sample flow rates called for in the theoretical curves, so we did not use them in these experiments.

Analytical Techniques. Numerous factors (e.g. inadequate theory, particle bounce, cross sensitivity, and unreproducible gas leakage) precluded accurate prediction of the true distributions of particles collected on the low-pressure impactor stages. Therefore we chose to verify the operation of the impactor by scanning electron microscopy (SEM). Particles from a 10% portion of each of the substrates the first MKV impactor sample obtained at plant A were dispersed into hexane by sonication, filtered on to Nuclepore filters, and sized from scanning electron micrographs by counting particles in discrete size ranges. A Quantimet image analyzer interfaced to the SEM sized the particles when the concentration was low enough to resolve individual particles. We were not able to obtain the distributions of particles on the back-up filters because the particles are small, and the Fluoropore filter is fibrous.

All impactor and filter samples were analyzed for up to 45 elements by instrumental neutron activation analysis (INAA) as described by Heft (17). Samples were irradiated simultaneously with standard flux monitors in the 3-MW Livermore pool reactor. The x-ray spectra of the radioactive species were taken with large-volume, high-resolution Ge(Li) spectrometer systems. The spectral data were transferred to a CDC 7600 computer and analyzed with the GAMANAL code (18), which incorporates a background-smoothing routine and fits the peaks with Gaussian and exponential functions.

Activation analysis with thermal neutrons is the method of choice because of the inherently high accuracy of the technique. Because matrix effects are virtually insignificant over the energy range of nuclear photons used in the analysis, the accuracy of the results is limited only by the counting statistics, the reproducibility of the solid angle impinging on the detector, and the accuracy of the standards. This is routinely within 5 to 10% for most elements determined. In comparing the mass of a single element in different samples or on sucessive impactor stages, however, uncertainity in the standard may be neglected. Further, because of the rather large distance between the sample and the detector, errors in positioning the sample are also small, and the limiting uncertainty becomes that of the counting statistics. The one-sigma uncertainty for Sc and As is often 2 and 5%, respectively, of the toal count. This high degree of precision permits interpretation of fine structure in the data that might otherwise be missed by other techniques.

Results and Discussion

Low-Pressure Impactor Data. Table I lists the count medians and geometric standard deviations from log-normal fits of the distributions obtained by SEM analyses of sample MKV-1. Listed also are the corresponding mass medians calculated from the count distributions.

The count-median diameters of 6 of the stages are below 1 μm, and the last 4 are well below the size range obtainable by inertial impaction at near-atmospheric pressure. Further, we operated the impactor at a somewhat higher final stage pressure than optimal (415 rather than 345 mm Hg), therefore, it is possible to obtain even smaller-size cuts. At this time, we have not thoroughly verified our SEM techniques for sizing aggregated particles smaller than 1 μm. Particles of the accumulation region are aggregates formed by coagulation of much smaller particles. Thus in sonically dispersing them, some bias towards smaller sizes might be expected if the aggregates are broken up. In practice, however, it is generally difficult to disperse submicrometer particles (especially particles smaller than 1 μm), because the relative surface forces are much stronger between particles of smaller size. Thus, it is just as likely that particles will disperse more nearly in the manner in which they came into the impactor, or that the SEM distributions are biased towards larger sizes. Ideally, the impactor will size particles into discrete, narrow intervals. This should reduce the effects of bias by sonic dispersion. Thus, we believe that the size distributions are qualitatively correct, cut caution must be exercised in using the data. We also recognize that liquid (or volatile solid) aerosols, such as sulfuric acid mists, cannot be sized by these techniques.

In Figure 2 we show the amount of Al, Fe, Sc, V, U, and Se in particles per log-size-interval of each impactor stage, per m^3 of gas plotted against the mass median diameters (mmd) of Table I. Note that in choosing the mmd and log-size interval for the filter, we assumed that the submicrometer distribution is log-normal and that all of the mass on the filter is contained in particles of diameters between 0.01 and 0.07 μm. These data suggest that the impactor intervals nicely bracket the accumulation mode that occurrs at 0.11 μm.

The submicrometer mode of all the 35 elements determined by INAA, occured at 0.11 μm except for those of Cr (Figure 3) and Mn (not shown) which occurred at 0.14 μm. Particles in this size range are so highly concentrated (10^{13} particles/cm^3 at 0.07 μm) that they coagulate very rapidly. Therefore, the shift in the modes of Cr and Mn may correspond to a temporal increment separating the formation of Cr and Mn aerosols from the other elements. Alternatively however, this may indicate that Cr and Mn aerosols are formed by a different mechanism than are the other elements. More careful analysis of the modes of the individual elements may provide insights into the trace-

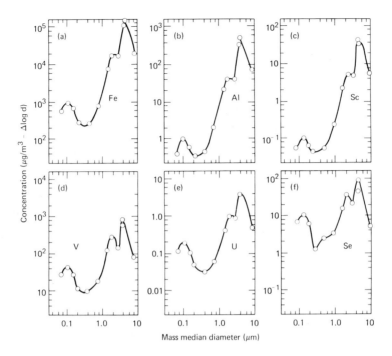

Figure 2. Concentration vs. size curves of Fe, Al, Sc, V, U, and Se in aerosol particles collected upstream of an electrostatic precipitator of a coal utility boiler.

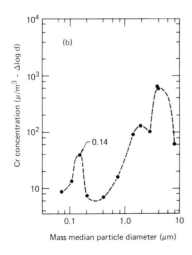

Figure 3. Profiles of concentration vs. particle size of Cr in particles collected upstream of an electrostatic precipitator of a coal utility boiler.

Table I. Log-normal distribution parameters and selected
analyses of particles on stages of impactor sample MKV-1[a].

Stage	CMD[b] (µm)	σg[c]	MMD[d] (µm)	Total mass (mg)	Al, mg	Sc, µg	As, µg
1	4.07	1.62	8.13	255	21±1	3080±20	1.3±0.3
2	2.70	1.48	3.79	558	39±2	6250±50	1.8±0.4
3	2.86	1.36	3.81	130	11.3±0.7	2270±20	1.4±0.4
4	1.87	1.43	2.73	35.2	2.9±0.4	686±5	0.72±0.07
5	1.14	1.48	1.82	28.0	2.6±0.1	639±4	0.75±0.04
6	0.76	1.58	1.43	16.5	1.8±0.3	371±3	0.45±0.4
7	0.50	1.46	0.77	--	0.22±0.03	53.6±0.5	0.133±0.05
8	0.184	1.66	0.40	--	0.053±0.005	13.5±0.3	0.088±0.002
9	0.15	1.44	0.22	--	0.034±0.003	8.7±0.3	0.104±0.004
10	0.10	1.36	0.14	--	0.041±0.004	9.1±0.4	0.167±0.003
11	0.077	1.42	0.11	--	0.103±0.007	21.1±0.4	0.436±0.006
12	--	--	--	4.54	0.146±0.009	41.3±0.7	1.03±0.02

[a]The impactor was operated at 7.84 slpm; the final stage
pressure was 415 mm Hg and stack gas temperature, 117°C.
[b]Count median diameter. [c]Geometric standard deviation.
[d]Mass median diameter. [e]The uncertainties reported are the
weighted means of the one-sigma uncertainties (from counting
statistics) of the multiple photopeaks used in the analysis as
described by Heft (17).

element combustion chemistry; that is to say, it may be possible
to discriminate between likely choices of chemical forms in
which the elements condense.

Chemical enrichment of aerosols. In Figure 4 we plot the
relative concentrations of W, V, U, and As in particles
collected in stack and at the ESP outlet location with the
8-stage impactors vs equivalent aerodynamic diameter (d_{50}).
Here the relative concentrations are expressed as the enrichment
factor (EF) rather than the weight/weight concentration (i.e.,
µg/g) because we cannot in every case accurately weigh the
particles on the stages. The EFs are approximately proportional
to the relative mass concentrations, and have the added
interpretive value of showing the chemical enhancement with

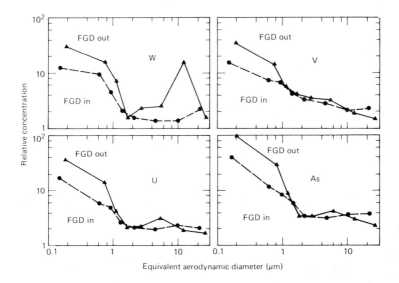

Figure 4. Enrichment factors (relative to Sc) vs. particle size curves for aerosols collected up- and downstream of the flue-gas desulfurization system show considerable concentration enhancement in the submicrometer size region.

respect to coal. The EFs are given by taking the ratios of the elements to that of Sc for the aerosol particles of each stage and dividing by the ratio of the same elements in pulverized coal.

In this data set, the concentrations of elements in submicrometer particles are clearly enhanced during passage through the FGD system. The concentrations of W, (Figure 4a), Cr, and Mn (Figure 5 a and b) show enhancement in both submicrometer- and super-micrometer-diameter particles. Selenium (Figure 6) was enhanced on particles of all sizes.

These findings are consistent with our earlier study of high-energy Venturi wet scrubber systems (12). It appears that high local acidity occurs when the aerosolized liquid droplets contact the flue gas, as evidenced by corrosion problems that often plague these devices. Note that while the pH of the liquid in the bottom of the scrubber is maintained at about 5.6, liquid aerosols extracted from the upper regions of the towers show much lower pH values. These conditions accelerate the dissolution of minor and trace elements on aerosol particle surfaces, thereby enriching the concentrations of the element in scrubber solutions. The bulk of the scrubbing solution is continuously injected into the spray sections by the recycle pumps, allowing considerable buildup of the leached substances in the scrubbing solution. On contact with the hot flue gas, the water evaporates from some of the aerosolized droplets. The enrichment may occurr by coagulation of the liquid droplets with fine particles, followed by evaporation of the liquid to leave a more highly chemically enriched particle of a somewhat larger, yet still quite small size. However, evaporation of water from the aerosolized droplets to form highly enriched submicrometer particles from the residue, may occur.

Finally, corrosion of internal metal surfaces, constituents of the absorbing and water-conditioning agents, and condensation of vapor components further contribute to the concentrations of Cr and Mn, Ca and P, and Se and S in the scrubbing solution and lead to their enrichment in fine aerosol emissions.

Estimates of the relative concentations of selected elements in fine particles collected upstream from the FGD system are listed in Table II. The values listed are some of the highest concentrations yet reported in the literature for fine coal- derived aerosols, especially those of V, Cr, and Zn, which lie in the tenth-percent range. Additional enhancements in the concentration of these elements resulting from FGD were about a factor of 2 for V, U, W, and As, a factor of 4 for Cr and Mn, and a factor of 15 for Se. If, as predicted by vapor-deposition models, the mass of there constituents resides in 0.02-μm-thick surface layers (10), then the surface concentration would be yet another 5 times greater, yielding concentrations of V, Zn, Cr, and Se in the range of 1 to 5%.

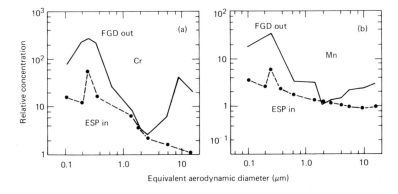

Figure 5. Enrichment factor vs. particle size curves for Cr and Mn show chemical enhancement of both large and smaller particles.

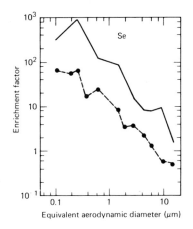

Figure 6. Enrichment factor vs. particle size curves for Se in aerosol particles sampled upstream of the Plant A cold-side electrostatic precipitator (– – –) and downstream of the Plant D flue-gas desulfurization system (———). Aerodynamic diameters of equivalent unit density spheres were obtained by adjusting the mmds of Table I for slip and particle density (assumed to be 2.4 g/cm³ for the particles collected).

Conclusions

We have operated the University of Washington MKV impactor
as a low-pressure impactor to provide for chemical analysis,
four discretely sized fly-ash fractions in the
sub-half-micrometer- diameter aerosol accumulation region.
Instrumental neutron activation analysis provided the
sensitivity to determine accurately the concentrations of 28
major, minor, and trace elements with sufficient precision to
reveal fine structure in the elemental distributions that might
be missed by techniques of lesser accuracy and precision.
We have further applied these techniques to investigate
the chemical modification of aerosols by a modern flue-gas
desulfurization system. This study confirms our earlier work
with a high-energy Venturi wet scrubber system, in which we
observed high chemical enrichment of aerosols from evaporative
processes.
In general, we feel that there are far too few studies of
the composition of particles in the submicrometer region. More
studies are needed to adequately characterize the human health
and environmental hazards associated with utility coal
combustion, and to more accurately determine their contribution
to urban pollutant inventories.

Table II. Estimates of the concentrations of selected elements
in submicrometer-diameter particles collected upstream of a
cold-side electrostatic precipitator, ppm.

Element	Filter	Stage 11	Stage 10
V	1900	1080	1170
Zn	1450	720	550
Cr	806	445	2400
As	230	140	130
Se	600	350	420
Mo	130	80	130
Co	160	20	66
Mn	160	80	20
U	107	68	68
W	76	49	63
Ni	-	-	780
Sb	82	47	48

REFERENCES

1. Ulrich, G.D., An Investigation of the Mechanism of Fly-Ash Formation in Coal-Fired Utility Boilers, Interim Report, US-ERDA FE-2205-1, May 28, 1976.
2. Flagan, R.C. and Friedlander, S.K. "Particle Formation in Pulverized Coal Combustion-A Review," presented at Symposium on Aerosol Science and Technology, Eight-Second National Meeting of the American Institute of Chemical Engineers, Atlantic City, N.J. 29 August -1 September 1976.
3. Ondov, J.M., Ragaini, R.C., and Biermann, A.H., Environ. Sci. Technol. 13, 946-953 (1979).
4. Ondov, J.M., Biermann, A.H., "Physical and Chemical Characterization of Aerosol Emissions from Coal-Fired Power Plants." in Environmental and Climatic Impact of Coal Utilization, J.J. Singh and A. Deepak, Eds., (Academic Press, New York) 1979.
5. Mims, C.A., Neville, M., Quann, R.J. and Sarofim, A.F., "Laboratory Studies of Trace Element Transformations During Coal Combustion," presented at the National 87th AICHE Meeting, Boston, 19-22 August (1979).
6. Neville, M., Quann R.J., Haynes, B.S., and Sarofim, A.F., "Vaporization and Condensation of Mineral Matter During Pulverized Coal Combustion," presented at the 18th International Symposium on Combustion, January (1980).
7. Davison, R.L., Natusch, D.F.S., Wallace, J.R., and Evans, C.A., Jr., Environ. Sci. Technol. 8, 1107-1113 (1974).
8. Smith, R.D., Campbell, J.A., Nielson, R.K. Environ. Sci. Technol. 13 593-558 (1979).
9. Biermann, A.H., and Ondov. J.M., Atmos. Environ. 14, 289-295 (1980).
10. Gladney, E.S., Small, J.A., Gordon, G.E., and Zoller W.H., Atmos. Environ. 10, 1071-1077 (1976).
11. Ondov, J.M., Ragaini, R.C., and Biermann, A.H., Atmos. Environ. 12, 1175-1185 (1978).
12. Ondov, J.M., Ragaini, R.C., and Biermann, A.H., Environ. Sci. Technol. 13, 598-607 (1979).
13. Ondov, J.M., Biermann, A.H., and Ralston, H.R., "Composition and Distribution Characteristics of Aerosols Emitted from a Coal-Utility Boiler Equipped with a Hot-Side Electrostatic Precipitator," presented to Annual American Chemical Society Meeting, Miami Beach, Sept 10-15 (1978).
14. MARK V University of Washington Source Test Cascade Impactor Pollution Control Systems Corporation, Renton, Washington.
15. Pilat, M.J., University of Washington, Seattle, private communication (1979).
16. Pilat, M.J., Powell, E.B., and Carr, R.C. "Submicron Particles Sizing with UW Mark 4 Cascade Impactor," in Proc. 70th Annual Meeting, Air Pollution Control Association, Vol. 4, 35.2 (1977).

17. Heft, R.E., paper presented at the Third International Conference on Nuclear Methods in Environmental Energy Research, Columbia, Mo., Oct 10-13, 1977.
18. Gunnink R., Niday, J.B., The GAMANAL Program, Lawrence Livermore Laboratory, Livermore, CA, UCRL-51061, Vols. I-III (1973).

RECEIVED March 25, 1981.

Sources and Fates of Atmospheric Polycyclic Aromatic Hydrocarbons

RONALD A. HITES

School of Public and Environmental Affairs and Department of Chemistry, Indiana University, 400 East Seventh Street, Bloomington, IN 47405

Polycyclic aromatic hydrocarbons (PAH) are produced by the combustion, under fuel rich conditions, of almost any fuel. Although a few PAH with vinylic bridges (such as acenaphthylene) are lost, most PAH are quite stable in the atmosphere and eventually accumulate in environmental sinks such as marine sediments. Spatial and historical measurements of PAH in sediments indicate that these compounds are stable, conservative markers of man's energy producing activities.

In 1775, Pursevil Pott first noted that the compounds associated with soot caused scrotal cancer in British chimney sweeps (1). Not having modern methods of instrumental analysis available to him, Pott was unable to specify the chemical structures of these compounds. It remained until 1933 before Cook et al. identified the exact structure of benzo[a]pyrene and demonstrated its carcinogenicity (2). Thus, polycyclic aromatic hydrocarbons (PAH) are one of the few groups of compounds which are known to be carcinogenic to man.

Although there are few chimney sweeps in business today, people are still exposed to considerable amounts of polycyclic aromatic hydrocarbons. Cigarette smoking, for example, is a major source of PAH (3); coke production also has high PAH emissions (4).

The exact synthetic chemistry which produces PAH in a fuel-rich flame is not well known, even today. It is clear, however, that PAH can be produced from almost any fuel burned under oxygen deficient conditions. Since soot is also formed under these conditions, PAH are almost always found associated with soot. As an example of the PAH assemblage produced by combustion systems, Figure 1 shows gas chromatographic mass spectrometry (GCMS) data for PAH produced by the combustion of kerosene (5). The structures of the major compounds are also given in Figure 1. We draw the reader's attention to a number of features of this PAH mix-

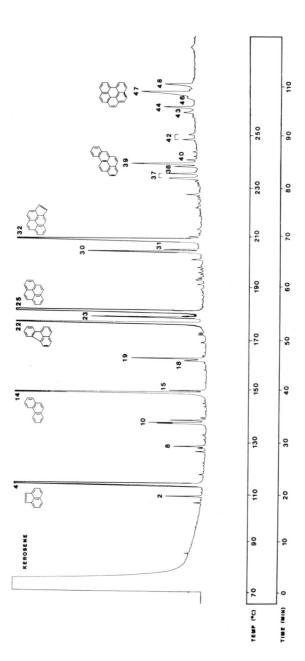

Biomedical Mass Spectrometry

Figure 1.   High-resolution gas chromatogram of PAH produced by the combustion of kerosene (5).
See Table I for peak identities.

Table I. PAH Identified by GCMS
(see Figures 1-3)

2	Biphenyl	32	Cyclopenta[cd]pyrene
4	Acenaphthylene	33	Benz[a]anthracene
8	Fluorene	34	Chrysene
10	$C_{14}H_8$[a]	35	Methylchrysene[c]
14	Phenanthrene	37	Benzofluoranthene
15	Anthracene	38	Benzo[e]pyrene
18	Methylphenanthrene	39	Benzo[a]pyrene
19	4H-cyclopenta[def]-	40	Perylene
	phenanthrene	42	$C_{21}H_{12}$ (unknown)
22	Fluoranthene	43	$C_{21}H_{12}$ (unknown)
23	Benz[e]acenaphthylene	44	Ideno[1,2,3-cd]pyrene
25	Pyrene	46	Dibenz[a,c]anthracene
27	Methylfluoranthene[b]	47	Benzo[ghi]perylene
30	Benzo[ghi]fluoranthene	48	Anthanthrene
31	$C_{18}H_{10}$ (unknown)		

[a] Probably cyclopent[bc or fg] acenaphthylene.

[b] Could be methylpyrene.

[c] Could be methylbenz[a]anthracene.

ture. First, fluoranthrene (peak 22) and pyrene (peak 25) are
present in about equal abundances. Second, the abundance of
phenanthrene far exceeds that of anthracene (peak 15), a less
stable compound. Third, benzo[a]pyrene (peak 39) is always found
with its noncarcinogenic isomer benzo[e]pyrene (peak 38).

A particularly interesting group of compounds in combustion
effluents are those with a vinylic bridge such as acenaphthylene
(peak 4) and cyclopenteno[cd]pyrene (peak 32). Peak 23, although
not labeled, has been positively identified as acephenanthrylene,
a compound which also has a vinylic bridge. We emphasize this
structural feature because of its chemical reactivity (compared
to the fully aromatic portions of the PAH). We shall see later
that this reactivity is important when considering the fate of
PAH in the atmosphere.

The PAH shown in Figure 1 are typical of those produced from
the combustion of various fuels (5). Without exception, the combustion of almost any fuel will produce the suite of compounds
shown in Figure 1. The relative abundances, however, can be substantially different depending on the temperature of combustion.
In fact, the relative abundance of the alkyl homologs of PAH, is
highly dependent on the temperature at which the fuel is burned
(6). Although Figure 1 shows very modest amounts of alkyl homologs (see the region between peaks 25 and 30), other fuels,
burned under other conditions, can show considerably greater a-
bundances of alkyl PAH. One can, in fact, use the relative abundance of the alkyl homologs to deduce the temperature at which
the fuel was burned.

Once the polycyclics are released from the combustion system,
presumably adsorbed on soot or fly ash, they are then exposed to
potential atmospheric degredation. A simple way in which to note
the relative degradation susceptibility of the various PAH is to
compare the GCMS data of the PAH coming from a combustion system
(see Figure 1) with the PAH profile of atmospheric particulates
(see Figure 2) (7). We see that those PAH without vinylic bridges are still prevalent, that the ratio of fluoranthrene to pyrene
is still about 1:1, and that the ratio of phenanthrene to anthracene is about 10:1. Those compounds with vinylic bridges [acena-
phthylene (peak 14), acephenanthrylene (peak 23), and cyclopenteno[cd]pyrene (peak 32)] have completely vanished from the PAH
mixture found in the atmosphere. Clearly, the increased chemical
reactivity of the relatively localized double bond found in these
compounds makes them susceptable to photolytic oxidation.

Assuming most PAH are stable in the atmosphere, which we
feel is an excellent assumption, we ask what happens to these
compounds after they are released from combustion systems
throughout the world. We suggest that PAH are transported to a-
quatic sediments either by direct airborne transport or by sediment resuspension and redeposition.

The basis for these arguments is extensive analyses of PAH
in sediments which we have carried out over the last several

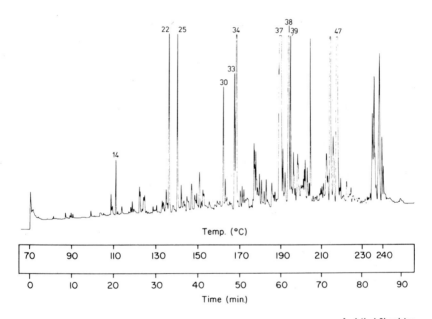

Analytical Chemistry

Figure 2. Capillary-column gas chromatogram of the total polynuclear hydrocarbon fraction of air-particulate matter (7). Conditions: 11 m × 0.26 mm –i.d. glass capillary coated with SE-52 methylphenylsilicone stationary phase; see Table I for peak identities.

years. The data shown in Figure 3 are among the first we obtained on sedimentary PAH (6, 8). This is a GCMS analysis of PAH
in the sediment of the Charles River, a rather polluted body of
water in Boston. By comparing this figure with the data shown in
Figure 2, one sees considerable resemblance. The ratios of the
major groups of compounds are the same. The PAH with vinylic
bridges are missing as they were in the atmosphere, and the alkyl
homologs are about as abundant as one might expect. We have obtained similar data, but in a more quantitative fashion, from
over 50 sediment samples from around the world (9). These data
indicate that PAH are ubiquitous and that they are found in almost all samples both near to and remote from urban areas. The
PAH pattern in all of these samples, even the most remote is similar to that shown in Figure 3.

    Even though the relative distribution remains constant, the
total level of PAH decreases dramatically with distance from urban centers. Figure 4 shows a plot of the total PAH abundance in
five marine sediment samples taken from Massachusetts Bay as a
function of distance from Boston (10). One can see that there is
a three order of magnitude decrease in the total abundance of PAH
within 100 kilometers of Boston. At that point, the total PAH
level is about 100 ppb; remarkably, this is what we see in almost
all other remote samples.

    Based on these and other measurements of PAH levels, we sug-
gest the following scenario for the transport of PAH. The various fuels which are burned in metropolitan areas produce airborne particulate matter (soot and fly ash) on which polycyclic
aromatic hydrocarbons are adsorbed. These particles are transported by the prevailing wind for distances which are a strong
function of the particle's diameter. We suggest that the long
range airborne transport of small particles accounts for PAH in
deep ocean sediments.

    Larger airborne particles will settle back onto the urban
area; rain then washes them from the streets and buildings. The
PAH in this urban run-off eventually accumulate in local sinks.
We suggest that these highly contaminated sediments are then
slowly transported by resuspension and currents to sea-ward locations where the sediments accumulate in basins or the deep o-
cean. The rapid decrease in PAH to a level of 160 ppb within 94
km of Boston (see Figure 4) indicates that this transport mode is
a rather short range effect (10).

    The stability of PAH is also apparent when one examines sed-
iment samples taken in such a way as to preserve the historical
record (11). This can be done by carefully coring sediments,
particularly at anoxic locations where there is little bioturba-
tion, segmenting the core into 2-4 cm sections, and analyzing
each section for PAH quantitatively. An example of such data is
shown in Figure 5; this represents a core from the Pettaquamscutt
River in Rhode Island, a highly anoxic basin (12). The total PAH
concentration ranges from 14,000 ppb near the sediment surface to

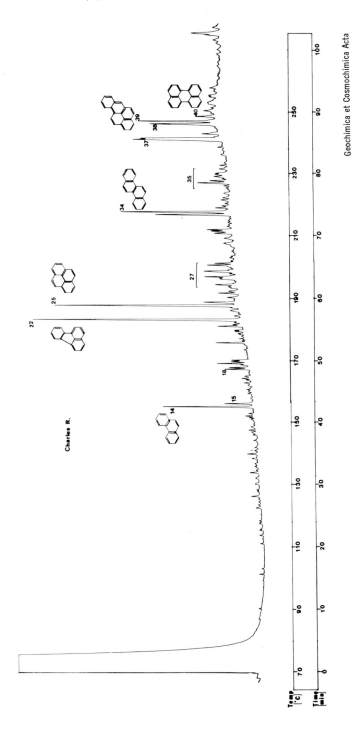

Figure 3. High-resolution gas chromatogram of PAH in Charles River sediment (6). See Table I for peak identities.

Geochimica et Cosmochimica Acta

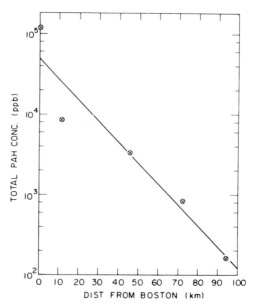

Geochimica et Cosmochimica Acta

Figure 4. Total PAH concentrations vs. distance from Boston for Massachusetts Bay samples (10)

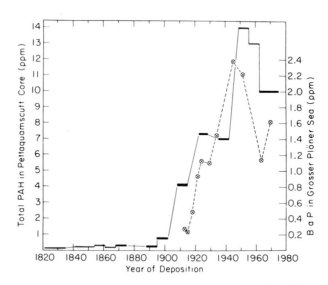

Geochimica et Cosmochimica Acta

Figure 5. Total PAH abundance in the various Pettaquamscutt River sediment core sections vs. date of deposition ((▬*), left scale); benzo[a]pyrene abundance in the Gosser Ploner Sea (14) vs. date of deposition (*⊗*, right scale) (12).*

less than 120 ppb at the core bottom. Despite the range of concentrations, the relative distribution of the PAH (excluding retene and perylene) is indicative of combustion. For example, the ratio of the $C_{16}H_{10}$ isomers (nonalkylated) to their monoalkyl homologs ($C_{17}H_{12}$) is 3.0 $\pm$ 0.4. In no case does this ratio become less than unity which would be expected if the source were direct fossil fuel contamination. The ratio of the $C_{16}H_{10}$ isomers to the $C_{18}H_{12}$ isomers is 2.7 $\pm$ 0.3, and the ratio of the $C_{14}H_{10}$ isomers to the $C_{20}H_{12}$ isomers is 0.46 $\pm$ 0.08. These ratios are consistent throughout the core and are indicative of combustion sources (6, 11, 12). We, therefore, conclude that combustion generated PAH predominate in all sections of the core.

Using the 3 mm/yr deposition rate reported by Goldberg et al. (13) a plot of total PAH (excluding retene and perylene) in the Pettaquamscutt core vs. year of deposition was developed (see Figure 5). For comparison, the benzo[a]pyrene data reported by Grimmer and Bohnke (14) for a core from the Grosser Ploner Sea are also plotted in Figure 5. The similarity between these two core profiles is quite remarkable. Both show rapid increases in PAH concentrations beginning around 1900. As discussed elsewhere (11, 12) this increase is certainly due to the heavy industrialization occurring at the turn of the century and the combustion associated with it.

A slight decrease in total PAH around 1930 is present in both cores (see Figure 5). It is intriguing to speculate that this reflects an event occurring both in Europe and New England at this time. The Depression could be such an event. During the Depression, the United States' total energy consumption decreased from 25 x 10^{15} BTU in 1929 to 18 x 10^{15} BTU in 1932 before resuming its increasing trend (15).

The Pettaquamscutt data are from a sufficiently deep core to allow us to assess the PAH burden prior to 1900. The PAH concentrations are at a low and constant level ($\sim$200 ppb) for the 50 yr previous to the turn of the century. This level may be indicative of PAH from natural combustion processes such as forest fires. Contributions from natural processes appear to be insignificant in areas or periods of high anthropogenic activity.

The decrease in PAH levels after 1950 is interesting and is similar to that observed at other locations (14). In our case, we think this reflects the change from coal to oil and natural gas as home heating fuels which occurred in the 1950's. During the period 1944-1961 the use of coal in the United States decreased by 40% while the use of oil and gas increased by 200% (15). Since combustion of coal usually produces more PAH than oil and gas (16), this change in fuel usage would result in a decrease in PAH production during the same period. We should point out that the possibility of returning to coal as a major energy source might, therefore, have a significant effect on man's input of PAH into the sedimentary environment.

In summary, we are suggesting that most PAH are a stable, conservative marker of man's energy producing activities and that this PAH record can be read with both spatial and historical resolution. In this context, PAH may be good, long-term indicators of air quality.

Acknowledgements

We are grateful to the National Science Foundation (Grant Numbers OCE-77-20252 and OCE-80-05997) and the Department of Energy (Grant Numbers EE-77-S-02-4267 and AC02-80EV-10449) for their support of our research on PAH. The author also thanks Milt Lee, Bob Laflamme, John Windsor, and John Farrington for their collaboration on the topics discussed here.

Literature Cited

1. Pott, P. "Chirurgical Observations"; Hawkes, Clarke, Collins: London, 1775.
2. Cook, J. W.; Hewett, C. L.; Hieger, I. J. Chem. Soc. 1933, 395.
3. Wynder, E. L.; Hoffman, D. "Tobacco and Tobacco Smoke. Studies in Experimental Carcinogenesis"; Academic Press: New York, 1967; p. 730.
4. Searl, T. D.; Cassidy, F. J.; King, W. H.; Benson, R. A. Anal. Chem. 1970, 42, 954.
5. Lee, M. L.; Prado, G. P.; Howard, J. B.; Hites, R. A. Biomed. Mass Spec. 1977, 4, 182.
6. Laflamme, R. E.; Hites, R. A. Geochim. Cosmichim. Acta 1978, 42, 289.
7. Lee, M. L.; Novotny, M.; Bartle, K. D. Anal. Chem. 1976, 48, 1566.
8. Hites, R. A.; Biemann, W. G. Adv. Chem. 1975, 147, 188.
9. Hites, R. A.; Laflamme, R. E.; Windsor, J. G. Jr. Adv. Chem. 1980, 185, 289.
10. Windsor, J. G. Jr.; Hites, R. A. Geochim. Cosmochim. Acta. 1979, 43, 27.
11. Hites, R. A.; Laflamme, R. E.; Farrington, J. W. Science. 1977, 198, 829.
12. Hites, R. A.; Laflamme, R. E.; Windsor, J. G. Jr.; Farrington, J. W.; Deuser, W. G. Geochim. Cosmochim. 1980, 44, 873.
13. Goldberg, E. D.; Gamble, E.; Griffin, J. J.; Koide, M. Estuarine Coastal Mar. Sci. 1977, 5, 549.
14. Grimmer, G.; Bohnke, H. Cancer Lett. 1975, 1, 75.
15. Hottle, H. C.; Howard, J. B. "New Energy Technology - Some Facts and Assessments"; MIT Press:Cambridge, 1971.
16. National Academy of Sciences "Particulate Polycyclic Organic Matter"; Nat. Acad. Sci.:Washington, D.C., 1972.

RECEIVED March 10, 1981.

Atmospheric Particulate Organic Matter

Multivariate Models for Identifying Sources and Estimating Their Contributions to the Ambient Aerosol

J. M. DAISEY and T. J. KNEIP

New York University Medical Center, Institute of Environmental Medicine, 550 First Avenue, New York, NY 10016

Multivariate regression models have been developed for apportioning the contributions of emission sources to airborne particulate organic matter. Weekly samples of respirable (<3.5 μm A.D., 50% cut) suspended particulate matter were collected in New York City from January, 1978 through August, 1979. The samples were analyzed for trace metals and sulfate as well as for three fractions of particulate organic matter (POM) using sequential extraction with cyclohexane (CYC), dichloromethane (DCM) and acetone (ACE). Factor analysis was used to identify the principal types of emission sources and select source tracers. Using the selected source tracers, models were developed of the form POM = a(V) + b(Pb) + - - -, where a and b are regression coefficients determined from ambient data adjusted to constant dispersion conditions. The models for CYC and ACE together, which constitute 90% of the POM, indicate that 40% (3.0 μg/m^3) of the mass was associated with oil-burning, 19% (1.4 μg/m^3) was from automotive and related sources and 15% (1.1 μg/m^3) was associated with soil-like particles. Comparison of the coefficients from the multiple regression analysis with available source emission data supports the validity of the models.

Although the ultimate source of much of particulate organic matter (POM) in the urban aerosol appears to be fossil fuel a specific knowledge of the amounts and classes of organic compounds contributed by various types of sources is lacking. Estimates of source contributions have been based on emission inventories which have been largely directed toward polycyclic aromatic hydrocarbons and/or benzo(a)pyrene. There has been very little work on the development of mathematical and statistical models for POM source identification and allocation (1). In view

of the success of Kleinman, et al.(2), in developing a multivariate statistical source apportionment model for TSP and potential
needs for regulation of airborne carcinogens, research on the
development of similar models for POM was initiated. The initial
objective of this work was to determine if such models could be
developed for POM. The results of this work to date are reported
here.

Methods

Sampling. Weekly samples of respirable ($\leq$3.5 µm, aerodynamic diameter, 50% cut) suspended particulate matter (RSP)
were collected on pre-ignited Gelman Type AE fiberglass filters
for organic analyses. Samples of total (TSP) and respirable suspended particulate matter were collected simultaneously on Gelman
Spectrograde filters for trace metal and sulfate analyses. The
sampling systems have been described (3, 4). The sampling site
is located on the roof of the residence hall (15th floor) at the
New York University Medical Center on East 30th Street, between
First Avenue and the F.D.R. Drive in New York City. Data used for
model development were from samples collected over the period
January, 1978 through August, 1979.

Analyses

A portion (10.2 cm x 17.8 cm) of each RSP sample was sequentially extracted in a Soxhlet apparatus with cyclohexane, dichloromethane and acetone, (8 hr. for each solvent) in the order
given. A more complete extraction of the organic compounds present in particulate matter is achieved and a partial separation
of the organic compounds into non-polar, moderately polar and
polar fractions is obtained by this method. The volume of each
extract was reduced to 10.0 ml using a rotary evaporator. The
samples were then stored in a freezer at -15°C until further
analysis. Weights of extracts were determined by weighing duplicate 100 µl aliquots of each, taken to dryness on a slide warmer
(40°C), on a Cahn Electrobalance.
   The average ambient concentrations of the cyclohexane-, di-
chloromethane- and acetone-soluble organic fractions from January,
1978 to August, 1979 were 2.9 µg/m^3, 1.0 µg/m^3 and 4.4 µg/m^3.
Respirable suspended particulate matter averaged 30.3 µg/m^3.
   Aliquots (2.54 cm x 20.3 cm) of the TSP and RSP samples
were analyzed for Pb, V, Mn and Cu using atomic absorption
spectrometry. The method has been described in detail previously
(3).
   Separate 2.54 cm x 20.3 cm aliquots of samples collected on
Spectrograde filters were analyzed for water-soluble sulfate as
described in a previous report (3). Sulfate data for RSP samples
were available only from October, 1978 since a third sampler had

to be used to obtain an RSP sample on a filter which was suitable
for sulfate determinations, i.e., Spectrograde.

Daily values of the 7 A.M. mixing height (MIXHT) and wind
speed (WDSPD) aloft at Fort Totten, N.Y. were obtained from the
National Weather Service. Weekly averages were computed for use
in the statistical analysis.

Weekly values of average temperature (AVETEMP) and of heating (HEATDD) and cooling (COOLDD) degree days were calculated
from the daily values measured in Central Park (in Manhattan)
obtained from the National Weather Service.

Trace metal, sulfate and organic species in RSP samples are
indicated with an "*" superscript. The symbols CYC, DCM and ACE
have been used for the cyclohexane-, dichloromethane-, and
acetone-soluble fractions of the RSP samples, respectively.

Dispersion Normalization. Atmospheric dispersion is greater
in winter than in summer in New York City and, in addition, varies
from year to year. Thus, for a constantly emitting source of
particles, atmospheric concentrations of TSP observed in winter
would be lower than in summer. In order to relate ambient concentrations of particulate species to their sources Kleinman,
et al. (5), suggested the use of a dispersion normalization technique based on the dispersion factor proposed by Holzworth (6).
Ambient concentrations of aerosol species are multiplied by the
ratio of the dispersion factor for the sampling period to an
average dispersion factor of 4200 m^2/sec.

$$DF = h \cdot s \qquad (1)$$

where DF is the dispersion factor, h is the mixing height, and s
is the wind speed aloft. Then

$$C_N = \frac{C \times DF}{4200} \qquad (2)$$

where C_N is the normalized concentration and C is the observed
concentration. The 7 A.M. mixing height and wind speed aloft,
measured at Fort Totten were used to compute the dispersion
factors. Dispersion normalization has been found to be a very
useful technique for gaining an understanding of both seasonal
and long-term trends in New York City (7, 8, 9).

Statistical Methods. The Statistical Package for the Social
Sciences (SPSS), version 8 (10, 11) was used for all statistical
analyses. All data were entered into a permanent file and verified prior to statistical analyses.

Factor analysis was used to identify principal types of
emission sources and to select elemental source tracers. As
factor analysis has been described in detail in the literature
(10-13), only a general description of the technique will be given

here. The classical factor analysis model used in this investigation is based on the assumptions that:

1. observed correlations between variables are due to some underlying structure in the data—source emissions and atmospheric reactions in this instance;

2. the observed variables are influenced by factors common to all variables and also by factors unique to each variable; the unique factors do not contribute to relations between variables and may be considered to include experimental errors in the measurements;

3. the factors are orthogonal;

4. the variables may be expressed as a set of n linear equations of the form:

$$x_{ji} = a_{j1}F_{1i} + a_{j2}F_{2i} + \ldots + a_{jp}F_{pi} + d_jU_{ji} \qquad (3)$$

where $j = 1, 2 \ldots n$ (number of variables)

x_{ji} = a variable in standardized form (eg., the difference between the concentration of an aerosol species and the mean concentration of that species divided by the standard deviation of the distribution of values for that species).

F_{pi} = the factors common to all of the variables.

U_{ji} = the unique factor.

a_{ji} = the factor loadings.

d_j = the standardized regression coefficient of variable j on unique factor U.

The values of the factor loadings, a_{ji}, rather than the factors themselves are of primary interest. The total variance of the variable accounted for the combination of all common factors is termed the communality of the variable and may be calculated from the factor loadings.

$$h_j^2 = a_{j1}^2 + a_{j2}^2 + a_{j3}^2 + \ldots \qquad (4)$$

where h_j^2 is the communality of the variable. The proportion of the variance not accounted for by the common factors is $1-h_j^2$.

The factor loadings also represent correlations between factors and variables and may be used to obtain the correlations between variables in the factor model. These calculated correlations should be close to the observed correlations.

Geometrically, the factors may be considered as orthogonal axes in n-dimensional space (n = number of variables). The variables are points in that space and the coordinates of the points are the loadings of the variables on the factors.

Principal factoring (R-type) with Varimax rotation of the factors (10) was employed in this investigation. Based on the results of the factor analysis, linear multiple regression models were developed in which organic species were related to concentrations of source tracer elements (predictor variables). These were of the form:

$$[POM] = k_1C_1 + k_2C_2 + \ldots k_iC_i + R \qquad (5)$$

where [POM] = the concentration of the organic species;
C_i = the concentrations of source tracer elements;
k_i = the regression coefficients determined by least equares analysis;
R = the residual, i.e., the portion of POM not associated with the modeled sources.

The constants k_i relate the source emissions to the ambient concentrations according to the relationship

$$k_i = \frac{\alpha_{ij} [POM]}{C_{ij}} \qquad (6)$$

where $[POM]_j$ = the concentration of a POM fraction or species in the emissions from the source j;
C_{ij} = the concentration of the tracer element in emissions from that source, and
α_{ij} = the coefficient of fractionation which describes changes in the ratio $[POM]_j/C_{ij}$ which may occur between the source and the ambient atmosphere as a result of physical and chemical processes.

The coefficients of equation (5) were determined by stepwise multiple regression in which the tracer element accounting for the greatest proportion of the variation of [POM] is used to find a first order, linear regression equation of the form [POM] = $k_1C_1 + R_1$, where R_1 is the residual. The partial correlation coefficients (correlation after allowance for correlation with the first tracer) of the remaining tracer elements are then examined. The C_i with the highest partial correlation coefficient with [POM] is selected next and a second regression equation of the form [POM] = $k_1C_1 + k_2C_2 + R_2$ is fitted. This process continues until all C_i which meet certain statistical criteria are included in the equation. In the first stage of model development the SPSS default criteria of F = 0.01 and T = 0.001 were employed. The values of F, T and also the r^2 which were obtained were always larger than these default criteria as factor analysis had been used to select appropriate tracer elements for the model. Listwise deletion of missing data was employed so that the model was fitted for those cases which included a complete set of variables.

Regression analysis assumes that all error components are
independent, have a mean of zero and have the same variance
throughout the range of POM values. Through an examination of
residuals, serious violations in these assumptions can usually be
detected. The standardized residuals for each of the fitted
models were plotted against the sequence of cases in the file
and this scatterplot was examined visually for any abnormalities
(10, 14).

Factor Analysis: Results and Discussion

Factor analysis was used as a qualitative and exploratory
technique to aid in the development of multiple regression models
for POM source apportionment. As a rather extensive data set was
available, different subsets of variables were analyzed by factor
analysis. A summary of the results is presented in Table I. The
most highly loaded variables for each factor are given for each
factor, along with the loadings of each variable. The loading
of the organic fractions on each factor are also given in all
instances in which the loading was greater than 0.1. In general,
depending upon the number of variables included (and, therefore,
the number of cases), 3 to 5 factors were obtained. In those instances in which meteorological variables were included, one
factor on which the variables MIXHT, WDSPD, and DF are highly
loaded was obtained. Previous investigations in this laboratory
have also indicated the importance of atmospheric dispersion in
New York City (7, 8, 9).

Source Tracers. The principal anthropogenic sources of
primary suspended particulate matter in New York City are transportation, fuel, oil combustion for power and space heating, and
incineration (2, 3, 15). From approximately November through
April, low sulfur (0.3%) fuel oil is burned for commercial and
residential space heating, and has been estimated to contribute
a substantial fraction of the TSP (2, 3, 15). Combustion of low-
sulfur fuel oil for power production is fairly constant throughout the year, and little or no coal is burned in the city. There
is no heavy industry, although small commercial operations such
as printing plants, electro-plating facilities, etc. may contribute to TSP (16).
In developing a multiple regression model for apportioning
sources of TSP in New York City, Kleinman, et al. (2) selected
Pb, Mn, Cu, V and SO_4 as tracers for automotive sources, soil-
related sources, incineration, oil-burning and secondary particulate matter, respectively. These were chosen on the basis of
the results of factor analysis and a qualitative knowledge of
the principal types of sources in New York City and the trace
metals present in emissions from these types of sources. Secondary TSP, automotive sources and soil resuspension were found to
be the principal sources of TSP in 1974 and 1975 (2).

The results of the factor analysis in the data set for 1978 through 1979 are in good agreement with the results reported by Kleinman, et al. (2). In general, V, $SO_4^=$, Pb, and Cu were found on separate and independent factors (cf. Nos. 1-3 in Table I). The element Mn* (RSP samples) was invariably associated with Pb*.

Mathematical relationships between tracer elements and organics will be affected by both physical and chemical factors. Thus, it was anticipated that relationships between POM fractions and source tracers would be more readily apparent if these species were all measured for samples of the same particle-size "cut" (1). This was confirmed by some preliminary investigations of relationships between POM in RSP samples and tracer elements in TSP samples in which few correlations were found. As POM is found largely in particles ≤ 3.5 μm aerodynamic diameter (17) and such particles penetrate most deeply into human lungs (18), POM was determined in RSP samples. Thus, it was important to demonstrate that the tracers (Pb, V, Cu, Mn, $SO_4^=$) in the RSP samples could be used as source tracers.

As can be seen in Table I, particularly in No. 1, the source tracers in the TSP and RSP samples tend to factor out together, with the exception of manganese. Manganese can be used as a soil tracer (2, 19, 20) in areas in which there are no other large sources of this element, such as steel manufacturing. Cooper and Watson (21) reported finding about 0.1% manganese in both the fine (<2 μm) and coarse (>2 μm) particle fractions, respectively, of urban dust in Portland and 0.2% and 0.085% in the fine and coarse fractions, respectively in continental dust. Thus, it should be possible to use this element in RSP samples as a soil tracer in New York City. However, during part of the period of this investigation, methylcyclopentadienyl tricarbonyl (MMT) was used as an octane booster for unleaded gasoline (8). As lead in automotive emissions has been found largely in the fine particle fraction (22, 23), it is reasonable to expect a similar size distribution pattern for manganese from automotive sources. This would explain the strong association between MN* and PB* in the factor analyses reported here. This probably also explains the association of MN (TSP samples) with several factors at low loadings, rather than with a single factor as the TSP samples would include the mass of respirable as well as coarse manganese. When the concentration of the tracers in coarse particles (calculated as the differences in concentration between the TSP and RSP samples) were included with the RSP data for factor analysis, a factor on which 20% of the total variance (No. 5, Table I) was loaded was obtained.

Particles in the coarse particle mode originate primarily from mechanical processes while those in the fine particle mode are produced by condensation and coagulation processes (24, 25). The work of many investigators (19, 20, 26, 27) indicates that manganese, particularly coarse particle manganese, is largely

Table I

SUMMARY OF RESULTS OF FACTOR ANALYSIS

VARIABLES INCLUDED	n[b].	FACTORS[a].				
		1	2	3	4	5
1. TSP trace metals and sulfate; RSP trace metals, Sulfate and organics; meteorological variables[c].	36	V (0.84) V* (0.82) HEATDD(0.71) CYC* (0.65) ACE* (0.21) (46.5%) oil burning/ space heating	SO4 (0.86) SO4* (0.85) COOLDD (0.71) DCM* (0.62) (24.4%) sulfate/ secondary	DF (0.92) MIXHT (0.81) WDSPD (0.75) (12.9%) meteorology	MN* (0.89) PB* (0.79) PB (0.62) ACE* (0.52) CYC* (0.25) (9%) auto/soil	CU (0.83) CU* (0.76) MN (0.53) DCM*(0.30) CYC*(0.15) (7.2%) incineration
2. TSP trace metals; RSP trace metals and organics	64	V (0.87) V* (0.84) CYC* (0.72) ACE* (0.35) (49.7%) oil burning	MN* (0.80) PB* (0.75) ACE* (0.45) CYC* (0.19) (23.6%) auto/soil	CU (0.90) CU* (0.71) (17%) incineration	MN (0.61) DCM* (0.54) CYC* (0.31) ACE* (0.13) (9.8%) soil	
3. RSP trace metals, organics, and sulfate; meteorological variables[c].	40	DF (0.94) WDSPD (0.84) MIXHT (0.84) HEATDD(0.76) TEMP (-0.76) (56.9%) meteorology	CYC* (0.79) V* (0.70) HEATDD(0.59) ACE* (0.23) (25.1%) oil burning/ space heating	PB* (0.82) MN* (0.78) ACE* (0.51) CYC* (0.13) (9.8%) auto/soil	SO4* (0.58) COOLDD(0.55) (8.2%) sulfate/ secondary	

Table I (continued)

SUMMARY OF RESULTS OF FACTOR ANALYSIS

VARIABLES INCLUDED	n[b]	FACTORS[a] 1	2	3	4	5
4. RSP trace metals, sulfate and organics	41	MN* (0.84) PB* (0.76) ACE* (0.45) CYC* (0.15) (51.6%) auto/soil	DCM* (0.97) SO4* (0.61) (30.3%) sulfate/secondary	V* (0.71) CYC* (0.63) ACE* (0.40) DCM (0.12) (18.1%) oil burning		
5. RSP trace metals, sulfate and organics; coarse particle trace metals[d]	38	MN* (0.83) PB* (0.77) ACE* (0.59) CYC* (0.22) (37.4%) auto/soil	V* (0.70) CYC* (0.70) ACE* (0.18) (31.4%) Oil burning	MN(C) (0.82) CU(C) (0.56) CU* (0.55) DCM* (0.40) ACE* (0.23) CYC* (0.12) (19.9%) soil/incineration	SO4* (0.80) DCM* (0.59) (11.2%) sulfate	

a. The most significant variables are given for each factor, along with the factor loadings of that variable; the percent of the total variance accounted for by each factor and source-related descriptors are also given for each factor.
b. n = number of cases (listwise deletion).
c. DF = dispersion factor; WDSPD = windspeed aloft; MIXHT = 7 A.M. mixing height; HEATDD = heating degree days; COOLDD = cooling degree days; TEMP = average temperature.
d. Coarse particle (C) = TSP-RSP.

derived from resuspended soil in many locations. A bimodal distribution has been observed for manganese in New York City (3,
28). The enrichment factor and particle mode distribution observed for this element (28) indicate that coarse particle
manganese is derived from windblown dust.

Based on these results and considerations, it was concluded
that Pb*, V*, Cu* and SO_4 could be used as source tracers for
automotive sources, oil-burning, incineration and secondary (or
sulfate-associated) sources, respectively. Mn*, however, should
probably be considered as a composite tracer of automotive
emissions and soil-related sources (resuspension of dust).
Coarse particle manganese (≥ 3.5 μm, aerodynamic diameter), MN(C),
was, therefore, used in the subsequent work as a resuspended soil
tracer.

Particulate Organic Fractions. Some very distinctive
patterns were seen for the associations of POM fractions and the
source-related factors. The cyclohexane-soluble fraction of POM
was invariably associated with V, heating degree days, and temperature (inversely), indicating that oil-burning (largely for
space heating) is the principal source of this fraction. The
fraction was also associated with the auto/soil and soil factors,
although at lower loadings (cf. No. 1-4, Table I).

In contrast to the CYC, the acetone-soluble fraction of POM
was not highly loaded on a single factor, but tended to be distributed among several factors (cf. Nos. 1, 2 and 4, Table I).
This suggested multiple sources, as did the correlation studies.
The strongest associations were with the factors identified as
auto/soil, oil-burning and incineration sources.

The dichloromethane-soluble fraction of POM showed a con-
sistant association with the sulfate factor. Previous summer
studies (29, 30) have indicated correlations between this fraction
and ozone. As sulfate and ozone are strongly correlated during
the summer months in the northeastern part of the U.S. (31),
there may be a stronger association between DCM and ozone than
DCM and sulfate. As of this date, the ozone data have been
obtained from the New York State Department of Environmental
Conservation but have not been entered into the computer files.
In addition, in view of the results of Schwartz, et al. (32), on
filter artifacts associated with the fraction, further studies
are warranted before developing a source apportionment model.
This fraction constitutes only about 10% of the total extractable
organic mass.

Multiple Regression Source Apportionment Models for Airborne
Particulate Organic Matter in New York City

Based on the results of the factor analysis, V*, PB*, CU*,
and MN(C) were chosen as source tracers for oil-burning, automobile emissions, incineration and resuspended soil, respectively.

The source apportionment models proposed for the CYC and ACE fractions were:

$$CYC^* = f(V^*, PB^*) \tag{7}$$
$$CYC = f(V^*, PB^*, MN(C)) \tag{8}$$
$$CYC = f(V^*, PB^*, MN(C), CU^*) \tag{9}$$
$$ACE^* = f(PB^*, V^*, MN(C)^*) \tag{10}$$
$$ACE^* = f(PB^*, V^*, \overline{MN(C)}, CU^*) \tag{11}$$

In the initial stages of this investigation, multiple regression models for ambient and dispersion normalized data (Eqn. 2) were compared. Summaries of the F values for each of the coefficients and their significance for the entry of each independent variable into the equation are presented in Table II for the models proposed in equations (7)-(11). An example of two equations obtained using ambient and dispersion-normalized data are presented in Table III. In every instance, better models were obtained with dispersion-normalized data. The coefficients for more source tracer variables were statistically significant when dispersion-normalized data were used. The equations had higher overall F values and r^2 values and, thus, accounted for a larger proportion of the variance; the standard errors of the coefficients and the residual terms were smaller.

The use of dispersion-normalized data is equivalent to adjusting all ambient concentrations to the same dispersion conditions and assuming that the remaining variations in concentrations are due to variations in source emissions. Although this is a logical approach conceptually, it is not known at present what uncertainties are associated with the use of a dispersion factor calculated from a 7 A.M. determination of mixing height and windspeed.

The 7 A.M. determination of mixing height appears to be a reasonably good estimate of the daily averages for the winter months but not for the summer (6, 33). From work in progress, the effect of using a 7 A.M. measurement rather than a daily average seems to be a secondary one, since the coefficients of the multiple regression coefficients reflect ratios of deviation of the variables from their means (14) and, therefore, change little when a daily average is used rather than a 7 A.M. value. The 7 A.M. data produce quite reasonable regression coefficients and estimates of source contributions (cf. discussion below).

In view of the large seasonal differences in atmospheric dispersion in New York City and the improved models obtained, dispersion-normalized data based on 7 A.M. measurements were used in this first stage of development for all of the source apportion tion models reported here. In order to keep the notation as simple as possible, this has not been indicated explicitly in the symbols used in the equations. Dispersion normalization, however, is implicit in all of the models reported.

Table II

COMPARISON OF AMBIENT AND DISPERSION NORMALIZED DATA IN MULTIPLE REGRESSION MODELS[a].

EQUATION NUMBER: (DEPENDENT VARIABLE)	AMBIENT DATA				DISPERSION NORMALIZED DATA			
	Step	Variable Entered	F to Enter Variable	p	Step	Variable Entered	F to Enter Variable	p
9 (CYC)	1	V*	40.3	<0.001	1	V*	269.7	<0.001
	2	MN(C)	1.1	0.295	2	MN(C)	20.6	<0.001
	3	PB*	0.8	0.372	3	CU*	5.4	0.024
	4	CU*	0.01	0.915	4	PB*	0.06	0.813
10 (ACE)	1	PB*	15.1	<0.001	1	PB*	97.0	<0.001
	2	V*	4.0	0.050	2	MN(C)	8.9	0.004
	3	CU*	0.9	0.338	3	V*	6.0	0.017
	4	MN(C)	0.8	0.380	4	CU*	0.4	0.541
11 (ACE)	1	PB*	15.1	<0.001	1	MN*	115.2	<0.001
	2	V*	4.0	0.05	2	V*	16.6	<0.001
	3	MN*	1.7	0.196	3	CU*	1.1	0.298
	4	CU*	0.8	0.386	4	PB*	0.2	0.686

a. n = 62 cases.

Table III

A COMPARISON OF SOURCE APPORTIONMENT MODELS
FROM AMBIENT AND DISPERSION NORMALIZED DATA[a].

AMBIENT DATA

$\text{CYC*} = 17.4\pm3.4\,[\text{V*}]+23\pm18\,[\text{MN(C)}]+0.35\pm0.38^{b\cdot}\,[\text{PB*}]+1.5\pm0.4$ (12)

$$n = 62, \; F = 14 \; (p < 0.001)$$
$$r^2 = 0.42$$

DISPERSION NORMALIZED DATA

$\text{CYC*} = 26.9\pm3.4\,[\text{V*}]+46\pm11\,[\text{MN(C)}]+0.3\pm0.3^{b\cdot}\,[\text{PB*}]+0.7\pm0.2$ (13)

$$n = 62, \; F = 126 \; (p < 0.001)$$
$$r^2 = 0.867$$

---

[a]. The standard error of the coefficient is given for each independent variable; the number of cases, overall F for the equation and its level of significance and r^2 are also given for each equation.

[b]. F value for coefficient is not statistically significant, i.e., at $p = 0.05$ level or better.

Source Apportionment Models for the Cyclohexane-Soluble
Fraction of Respirable Suspended Particulate Matter. Stepwise
multiple regression analysis was used to determine the coefficients of the source tracers for the models proposed for CYC*
in equations (7)-(9). These models are shown in Table IV. As
expected from the factor analyses, the coefficient for V*, accounting for the greatest proportion of the variance of CYC*, was
fitted first into the equation. Equation (14) was the simplest
and the F value was slightly higher than for equations (15) and
(16). In addition, as will be discussed later in this paper, the
coefficient for PB* was in reasonable agreement with the ratio of
CYC*/PB* for samples collected in the Allegheny Tunnel.
     When MN(C) is included as a variable the coefficient for PB*
became statistically insignificant (p = 0.36). The coefficient
for PB* in equation (15) was a factor of two smaller than in
equation (16); however, this is not a significant difference in
view of the uncertainties of the coefficients, particularly in
equation (16). If additional variables (CU* or MN*) were included in the multiple regression analysis, these were placed in the
equations ahead of PB*; the coefficient for PB* was not statistically significant in any of these equations and often negative.
This can be seen in equation (16). If MN* was used as a mixed
auto/soil tracer, equation (17) was obtained. Although this does
not allow separate estimates of the contributions of automobiles
and soil resuspension, the particle-size of the tracer is more
appropriate.
     The contributions of each type of source were calculated from
each equation using the dispersion normalized average concentrations of the trace metals. The results are presented in Table
V. In terms of the major source, oil-burning, there is very
little difference in the estimated contributions from one equation
to the next. All of the equations indicate that oil burning contributed about 50% (1.5-1.7 $\mu g/m^3$) of the CYC fraction during
this period (January, 1978 through August, 1979). Transportation,
largely automotive, was estimated to contribute 0.57 $\mu g/m^3$ (17.5%)
using equation (14). As additional variables were included in the
equations, the estimated contribution diminished. Equations (15)
and (16) suggest that about 20% of the CYC fraction originated
from resuspended soil. Equation (16) suggests 12% of CYC from
incineration during this period.
     Using equation (17) in which MN* is used as a mixed auto/
soil tracer, these two sources in combination were estimated to
contribute 0.9 $\mu g/m^3$ (27%) of CYC* during the period of this investigation. This is consistent with the results of equations
(15) and (16).
     A residual of 20-30% (0.65-0.99 $\mu g/m^3$) could not be attri-
buted to any of the sources. Seasonal differences in temperature
undoubtedly affect the vapor/particulate equilibrium for the non-
polar organics in this fraction and, thus, contribute to the unexplained portion of CYC*. Attempts to take this into account

Table IV

SOURCE APPORTIONMENT MODELS FOR CYCLOHEXANE-SOLUBLE
PARTICULATE ORGANIC MATTER[a].

$$CYC* = 29\pm4[V*]+0.65\pm0.35[PB*]+1.0\pm0.2 \tag{14}$$

$$n = 62, \; F = 142 \; (p < 0.001)$$
$$r^2 = 0.828$$

$$CYC* = 27\pm3[V*]+46\pm11[MN(C)]+0.3^{b.}\pm0.3[PB*]+0.7\pm0.2 \tag{15}$$

$$n = 62, \; F = 126 \; (p < 0.001)$$
$$r^2 = 0.867$$

$$CYC* = 26\pm3[V*]+42\pm11[MN(C)]+16\pm8[CU*]+0.08^{b.}\pm0.3[PB*]+0.65\pm0.2 \tag{16}$$

$$n = 62, \; F = 101 \; (p < 0.001)$$
$$r^2 = 0.877$$

$$CYC* = 30\pm3[V*]+99\pm32[MN*]+0.7\pm0.2 \tag{17}$$

$$n = 66, \; F = 179 \; (p < 0.001)$$
$$r^2 = 0.850$$

[a]. Concentrations in $\mu g/m^3$; all data dispersion normalized. Values of F and r^2 are given for the overall equation and the standard error of each coefficient is given.

[b]. Regression coefficient is not statistically significant at $p \leq 0.05$ level.

Table V

CALCULATED SOURCE CONTRIBUTIONS TO THE CYCLOHEXANE-SOLUBLE

FRACTION OF AIRBORNE PARTICULATE MATTER, $\mu g/m^3$[a].

FROM EQN. NO.	Oil-burning V*	Transportation PB*	Soil Resuspension MN(C)	Incineration CU*	Residual
14	1.69 (52%)	0.57 (17.5%)	–	–	0.99 (30.5%)
15	1.57 (48.1%)	0.26[b] (8%)	0.70 (21.5%)	–	0.73 (22.4%)
16	1.49 (45.8%)	0.07[b] (2.2%)	0.64 (19.7%)	0.40 (12.3%)	0.65 (20%)
17	1.73 (52%)	0.90[c] (27%)	–	–	0.70 (21%)

[a] Concentrations are dispersion normalized to 4200 m^2/sec; the percent contribution of each source is given in parentheses.

[b] Coefficient for PB* not statistically significant in this equation.

[c] MN* used as a mixed auto/soil tracer.

explicitly in the model have been unsuccessful to date. There
are probably unidentified sources of non-polar organics in New
York City, as well, including transport from areas outside of
the city (1).

The most appropriate models for CYC* is probably equation
(14) as this is the simplest and is most closely related to the
results of the factor analysis which indicated that the variances
in CYC* were most closely related to those in V* and PB*. The
inclusion of coarse particle manganese as a soil tracer diminished the significance of the coefficient of PB* and the contribution of automotive sources. Ideally, MN* would be used as a
tracer for resuspended soil but interferences from the use of MMT
as a fuel additive during part of the period in which this data
were collected make this a mixed source tracer for the contributions of automobiles and soil resuspension.

Source Apportionment Models for the Acetone-Soluble Fraction
of Respirable Suspended Particulate Matter. The multiple regression models for ACE*, based on equations (10) and (11) are presented in Table VI. The F values for all of the coefficients in
equation (18) are statistically significant (p $\leq$ 0.017). In contrast to the case for CYC*, there was no difficulty in obtaining
a coefficient for PB* for the ACE* fraction. This is probably
due to the fact that automotive sources contribute more substantially to this fraction than they do to the CYC* fraction. Copper
was included as a tracer in equation (19); however, the F value
for the coefficient was not statistically significant (p = 0.54)
and the change in r^2 is only 0.002. Thus, equation (18) was used
to estimate the contributions of the sources to the dispersion
normalized and ambient concentrations of ACE. The results of
these calculations are presented in Table VII. Thirty-six percent
of the ACE* fraction appears to originate from automotive sources
while oil burning and soil resuspension each account for about
20%.

Using equation (20), with MN* used as a mixed tracer to calculate source contributions, 2.64 $\mu g/m^3$ (53%) of ACE* was estimated to originate from automotive sources plus soil resuspension,
in good agreement with the calculations based on equation (18).

Approximately 80% of the mass of ACE* fraction can be accounted for using equations (18) or (19). This compares favorably
with the TSP model developed previously in this laboratory (2)
which accounted for about 75% of the TSP in New York City.

Model Verification. There are several ways in which the
models described here may be verified:

1. by comparison with source emission data;
2. by comparison of the multiple regression coefficients
   to ratios of organics to tracer elements in source
   emissions;
3. by using the model to predict concentrations of organics
   at another site.

Table VI

SOURCE APPORTIONMENT MODELS FOR THE ACETONE-SOLUBLE
FRACTION OF RESPIRABLE PARTICULATE MATTER[a].

$ACE* = 2.0\pm0.7[PB*]+70\pm26[MN(C)]+19\pm8[V*]+1.0\pm0.4$ \hfill (18)

$$n = 62, F = 45 \ (p < 0.001)$$
$$r^2 = 0.699$$

$ACE* = 1.9\pm0.8[PB*]+66\pm26[MN(C)]+19\pm8[V*]+12\pm19[CU*]+0.9\pm0.4$ \hfill (19)

$$n = 62, F = 33 \ (p < 0.001)$$
$$r^2 = 0.701$$

$ACE* = 293\pm64[MN*]+25\pm6[V*]+0.84\pm0.43$ \hfill (20)

$$n = 66, F = 79.8 \ (p < 0.001)$$
$$r^2 = 0.72$$

---

[a]. Concentrations in $\mu g/m^3$; all data dispersion normalized. The
values of F and r^2 are given for the overall equation and the
standard error of each coefficient is given.

Table VII

ESTIMATES OF SOURCE CONTRIBUTIONS TO THE DISPERSION NORMALIZED
AND AMBIENT CONCENTRATIONS OF THE ACETONE-SOLUBLE FRACTION
OF RESPIRABLE SUSPENDED PARTICULATE MATTER[a].

SOURCE	TRACER	DISPERSION NORMALIZED, $\mu g/m^3$	CONTRIBUTIONS AMBIENT, $\mu g/m^3$	%
Automobiles	PB*	1.79	1.64	36.0
Soil Resuspension	MN(C)	1.06	0.97	21.4
Oil Burning	V*	1.13	1.04	22.8
Residual	–	1.0	0.90	19.8

[a]. Based on equation (18).

At present there are no source emission data available for
POM in New York City or even for TSP. There is an estimate of
primary TSP emissions for 1977 based on projections of energy
usage (15). If this estimate is used, with an additional 15%
TSP added for soil resuspension (2), then oil-burning, transportation, soil resuspension and other sources (such as incineration)
can be estimated as contributing 57%, 24%, 15% and 4% of the primary TSP in New York City, respectively. The relative contributions of these sources to POM, i.e., the sum of CYC* and ACE*,
from equations (14) and (18) are shown in Table VIII. Assuming
that particulate organic matter is approximately proportional to
TSP, the agreement in estimated relative source contributions is
quite reasonable.
   The coefficients of the source tracers in the multiple regression equations can also be used for verification. From equation (6), these coefficients are approximately equal to the ratios
of organics to tracer for each source. The coefficients and
source ratios will not be identical due to the changes which may
occur between the source and the atmosphere. In addition, the
coefficient in the multiple regression equation is an average
ratio for all sources of a particular type for a given period.
Measurements of source emissions are generally for a single source
at a single point in time.
   Samples of TSP, collected in the Allegheny Tunnel in 1978
under different mixes of diesel and automotive traffic*, were
analyzed for CYC* and for PB*. The ratios of CYC*/PB* were then
plotted versus % diesel traffic and the line obtained was ex-
trapolated to the intercept, i.e., no diesel traffic, to determine the ratio for automobiles. The ratio obtained was 1.6. From
equation (14), the PB* coefficient, for the period from January,
1978 through August, 1979, was 0.65 ± 0.35, in good agreement with
the ratio for the Allegheny Tunnel. If only the data for May
through September, during which there is no space heating (and no
overwhelming influence of V), are used, an equation relating CYC*
to PB* can be obtained for an extended summer season:

$$\text{CYC*} = 72 \pm 18\,[\text{MN(C)}] + 2.2 \pm 0.7\,[\text{PB*}] + 0.6 \pm 0.3 \qquad (21)$$

$$n = 21, \quad F = 16.2 \ (p < 0.001)$$
$$r^2 = 0.644$$

   The F values for the overall equation and for each of the coefficient, were all statistically significant at the $p \leq 0.01$ level.

From equation (21)

$$\alpha_{ij} \frac{\text{CYC*}}{\text{PB*}} = 2.2 \pm 0.7 \qquad (22)$$

*Provided by Dr. William Pierson, Ford Motor Co., Dearborn, MI.

Table VIII

COMPARISON OF ESTIMATES OF RELATIVE SOURCE CONTRIBUTIONS
FOR POM[a]. and TSP[b].

| Source | ESTIMATED CONTRIBUTIONS TO | |
	POM[a]. (from models)	TSP[b].
Oil-burning	40%	57%
Transportation	19%	24%
Soil Resuspension	15%	15%
Other	–	4%
Residual	27%	–

[a]. POM = CYC* + ACE*

[b]. Based on Jones, et al. (15) with an additional 15% contribution from soil resuspension added.

for transportation sources in New York City. This also includes
a low percent diesel traffic. The coefficient of 2.2 from equation (21) is in very good agreement with the ratio of CYC*/PB*
(1.6) obtained from the Allegheny Tunnel samples.

Data on urban soil from the Portland Aerosol Study (21) were
used to obtain an order of magnitude comparison of source ratios
to the coefficients of MN(C) in the models. If volatile carbon
(21) is assumed to be approximately equal to extractable organic
matter (this study) and using a 1:1 ratio for coarse to fine
particle mass in New York City (based on our unpublished data),
then a ratio of extractable organic matter to MN(C) of 118 can be
estimated for urban soil. The coefficients for MN(C) in the
models were 46±11 [equation (16)] and 70±26 [equation (19)] for
CYC* and ACE*, respectively. This is quite reasonable agreement
in view of the approximations made to obtain a ratio for the soil
source.

In summary:

1. the coefficients obtained for PB* in the regression
   equations for CYC* are in very good agreement with the
   source emission data for the Allegheny Tunnel;
2. the coefficients for MN(C) are of the same order of magnitude as the soil source ratios estimated from literature data;
3. the estimates of the contributions of various type of
   sources to ambient concentrations of CYC* and ACE* are
   consistent with available estimates of source contributions of TSP, assuming that organics are proportional to
   TSP.

As the data collected for the Queens sampling station become
available it will be possible to test the models in another location, i.e., to predict concentrations of CYC* and ACE* for the
Queens station. Efforts to obtain data for emission sources for
further verification of the models are also underway.

Summary and Conclusions

Multivariate source apportionment models have been developed
for two fractions of respirable particulate organic matter which
together constitute about 90% of the total organic solvent-ex-
tractable mass. The independent variables used for developing
the models were trace metals, water-soluble sulfate and meteorological variables. Two of the three POM fractions extracted
sequentially with cyclohexane (CYC), dichloromethane (DCM) and
acetone (ACE) were used as individual dependent variables.

The data were analyzed first by factor analysis. Depending
upon the number of variables included in the analysis, three to
five factors were obtained which could be associated with emission
sources or with meteorology. The source tracers V*, Pb*, Cu* and
SO_4* were generally found on independent factors and were the
variables most highly loaded on those factors. The CYC fraction

was most strongly associated with vanadium, a tracer for oil-burning sources, while the ACE fraction was associated with automotive
and auto/soil source tracers as well as vanadium. A meteorological factor, which included the variables wind speed and mixing
height, was found to be vary significant, confirming previous
work by our laboratory.

Using the tracers selected from the factor analysis, models
were developed of the form CYC* = a(V*) + b(Pb*) + - - - - - + f,
where the coefficients a and b and the constant f were determined
by multiple regression analysis. By using dispersion normalized
concentrations, rather than ambient concentrations, the statistical significance of the models was greatly improved. This is
equivalent to adjusting all ambient concentrations to the same
dispersion conditions and assuming that the remaining week-to-
week variations in the concentrations are due to those in source
emissions.

The models for the CYC and ACE fractions together indicate
that 40% (3.0 $\mu g/m^3$) of these two fractions in respirable
particles in New York City was associated with oil burning, 19%
(1.4 $\mu g/m^3$) was from automotive and related sources and 15%
(1.1 $\mu g/m^3$) was associated with soil-like particles.

Comparisons of the regression coefficients of the source
tracer elements with available source emission data, as well as
comparisons with estimated source emission data for total suspended particulate matter, provide evidence of the validity of the
models.

These results have demonstrated that it is possible to
develop source-receptor apportionment models for airborne
particulate organic matter using ambient data. The utility of
incorporating meteorological data in such models has also been
shown. Further development of such models should include efforts
to verify more fully the validity of these models through comparisons with source emission data. Additional evidence of their
validity should be sought by testing such models as predictive
tools in other sites. It may also be possible to explicitly include factors such as atmospheric degradation and the effects of
temperature on vapor-particulate equilibria in future models and,
thus, learn more about the dynamics of organic compounds in the
atmosphere.

Acknowledgements

    We are grateful to Dr. W. R. Pierson of the Ford Motor Com-
pany for providing samples of suspended particulate matter which
were collected in the Allegheny Tunnel. We thank Dr. M. T.
Kleinman, University of Southern California, for his comments
and suggestions.

    This work was supported by Grant No. RP-1222 of the Electric
Power Research Institute and is part of a center program supported
by the National Institute of Environmental Health Sciences, Grant
No. ES00260, and The National Cancer Institute, Grant No. CA 13343.

Literature Cited

1. Daisey, J. M., M. A. Leyko and T. J. Kneip. Source identi-
    fication and allocation of PAH compounds in the New York City
    aerosol: Methods and applications. In: Polynuclear Aroma-
    tic Hydrocarbons, P. W. Jones and P. Leber, eds. Ann Arbor
    Science Publishers, Inc., Ann Arbor, Michigan. 201-215
    (1979).

2. Kleinman, M. T., B. S. Pasternack, M. Eisenbud and
    T. J. Kneip. Identifying and Estimating the Relative im-
    portance of Sources of Airborne Particulates. Environ. Sci.
    and Tech. 14: 62-65 (1980).

3. Eisenbud, M. and T. J. Kneip. Trace Metals in Urban
    Aerosols. EPRI Report No. RP-117. (1975), NTIS #Pb-248-324.
    (1976).

4. Kneip, T. J., M. Lippmann, F. Mukai and J. M. Daisey. 1980.
    Trace Organic Compounds in the New York City Atmpsohere.
    Final Report on RP-1222 to the Electric Power Research In-
    stitute, Palo Alto, Ca. In preparation.

5. Kleinman, M. T., T. J. Kneip and M. Eisenbud. Seasonal
    patterns of airborne particulate concentrations in New York
    City. Atmos. Environ. 10: 9-11 (1976).

6. Holzworth, G. C. Mixing depths, wind speeds, and air pollu-
    tion potential for selected localities in the U.S. J. Appl.
    Meteorol. 6: 1039 (1967).

7. Kleinman, M. T., D. M. Bernstein and T. J. Kneip. An ap-
    parent affect of the oil embargo on total suspended partic-
    ulate matter and vanadium in New York City air. J. Air
    Pollut. Contr. Assoc. 27: 65-67 (1977).

8. Eisenbud, M. Levels of exposure to sulfur oxides and partic-
    ulates in New York City and their sources. Bull. N.Y. Acad.
    Med. 54: 991-1011 (1978).

9. Lioy, P. J., R. P. Mallon and T. J. Kneip. Long-term trends
    in total suspended particulates, vanadium, manganese and
    lead at near street level and elevated sites in New York
    City. J. Air Pollut. Contr. Assoc. 30: 153-156 (1980).

10. Nie, N. H., C. H. Hull, J. G. Jenkius, K. Steinbrenner and
    D. H. Beat. SPSS Statistical Package for the Social Sciences,
    2nd ed. McGraw-Hill Book Co., New York, N.Y. (1975).

11. Hull, C. H. and N. H. Nie. Editors. SPSS Update. New Procedures and Facilities for Releases 7 and 8. McGraw-Hill Book Co., New York, N.Y. (1979).

12. Harman, H. H. Modern Factor Analysis, 3rd ed. University of Chicago Press, Chicago, Ill. (1976).

13. Malinowski, E. R. and D. G. Howery. Factor Analysis in Chemistry. John Wiley and Sons, New York, N.Y. (1980).

14. Draper, N. R. and H. Smith. Applied Regression Analysis. John Wiley and Sons, New York, N.Y. (1966).

15. Jones, H. G. M., P. F. Palmedo, R. Nathans, et al. Energy Supply and Demand in the New York City Region. Report No. BNL 19493. Energy Systems Analyses Group, Brookhaven National Laboratory, Upton, N.Y. 11973. Urban and Policy Sciences Program, State University of New York, Stony Brook, N.Y. 11290 (1974).

16. Kleinman, M. T. The apportionment of sources of airborne particulate matter. Doctoral dissertation, New York University, New York, N.Y. (1977).

17. Daisey, J. M. Organic Compounds in Urban Aerosols. Ann. N.Y. Acad. Sci. 50-69 (1980).

18. Lippmann, M. "Size-selective sampling for inhalation hazard evaluations," In Fine Particles, Aerosol Generation, Measurement, Sampling and Analysis. B. Y. H. Liu, ed. Acedemic Press, Inc., New York, N.Y. (1976).

19. Rahn, K. A. Sources of trace elements in aerosols, an approach to clean air. Doctoral dissertation, University of Michigan, Ann Arbor, Michigan. (1971).

20. Miller, M. S., S. K. Friedlander and G. M. Hidy. A chemical element balance for the Pasadena aerosol. J. Colloid. Interface Sci., 39: 165-176 (1972).

21. Cooper, J. A. and J. G. Watson. Portland Aerosol Characterization Study (PACS). Application of Chemical Mass Balance Methods to the Identification of Major Aerosol Sources in the Portland Airshod. Final report Summary. Prepared for the Portland Air Quality Maintenance Area Advisory Committee and the Oregon Department of Environmental Quality. April 23, 1979.

22. Dzubay, T. G., R. K. Stevens and L. W. Richards. Composition of aerosols over Los Angeles Freeways. Atmos. Environ. 13: 653-659 (1979).

23. Little, P. and R. D. Wiffen. Emission and deposition of lead from motor exhausts. II. Airborne concentration, particle size and deposition of lead near motorways. Atmos. Environ. 12: 1331-1341 (1978).

24. Whitby, K. T., R. B. Husar and B. Y. H. Liu. The aerosol size distribution of Los Angeles smog. J. Colloid Inter. Sci. 39: 177-204 (1973).

25. Lippmann, M. Size distribution in urban aerosol. Ann. N.Y. Acad. Sci. 338: 1-12 (1980).

26. Hammerle, R. H. and W. R. Pierson. Sources and elemental composition of aerosol in Pasadena, Calif. An energy-dispersive x-ray fluorescence. Environ. Sci. Tech. 12: 1058-1068 (1975).

27. Dzubay, T. G. Chemical element balance method applied to dichotomous sampler data. Ann. N.Y. Acad. Sci. 338: 126-144 (1980).

28. Bernstein, D. M. and K. A. Rahn. New York Summer Aerosol Study; Trace Element Concentrations as a Function of Particle Size. Ann. N.Y. Acad. Sci. 322: 87-97 (1980).

29. Daisey, J. M., M. A. Leyko, M. T. Kleinman and E. Hoffman. The nature of the organic fraction of the New York summer aerosol. Ann. N.Y. Acad. Sci. 322: 125-141 (1979).

30. Daisey, J. M., R. J. Hershman and T. J. Kneip. Seasonal variations in ambient levels of particulate organic matter in New York City. Paper presented at the Symposium on Atmospheric Aerosols, 178th National Meeting of the American Chemical Society, Washington, D.C., Sept. 9-14, 1979.

31. Lioy, P. J., T. J. Kneip, M. Lippmann, P. Mallon and P. Samson. Interim Report on Project HOST. Report to U.S. Environmental Protection Agency, August, 1980.

32. Schwartz, G., J. M. Daisey and P. J. Lioy. Effect of sampling duration on the concentration of particulate organics collected on glass fiber filters. Am. Ind. Hyg. Assoc. J., In press.

33. Hanna, S. R. Diurnal Variation of the Stability Factor in the Simple ATDL Urban Dispersion Model. J. Air Pollut. Contr. Assoc. 28: 147-150 (1978).

RECEIVED May 6, 1981.

An Automated Thermal–Optical Method for the Analysis of Carbonaceous Aerosol

RICHARD L. JOHNSON, JITENDRA J. SHAH, ROBERT A. CARY, and JAMES J. HUNTZICKER

Department of Environmental Science, Oregon Graduate Center, 19600 N.W. Walker Road, Beaverton, OR 97006

An instrument employing both thermal and optical measurements has been developed for the analysis of organic and elemental carbonaceous aerosol collected on glass or quartz fiber filters. The technique involves volatilization of organic carbon from the filter under conditions where the elemental carbon remains. The volatilized carbon is oxidized to CO_2, reduced to CH_4, and measured by a flame ionization detector. Elemental carbon is subsequently oxidized to CO_2 and measured. The reflectance of the filter is continuously monitored throughout the analysis by a helium–neon laser system. During the organic analysis some of the organic carbon is pyrolytically converted to elemental carbon; this results in a decrease in the filter reflectance. Correction for the pyrolytic production of elemental carbon is achieved by measuring the amount of elemental carbon oxidation necessary to return the filter reflectance to its initial value. The instrument is completely automated and is under the control of a microprocessor system. It has been evaluated with respect to model compounds, typical source mixtures (e.g., auto exhaust aerosol), and ambient samples.

Although carbon has long been recognized as an important constituent of ambient aerosols, the analysis of carbon in its many molecular forms has presented formidable obstacles. An approach taken by many investigators (1-12) has been to separate aerosol carbon into organic, elemental, and carbonate classes. However, at the present time only carbonate carbon has an unequivocal analytical definition. Speciation between organic and elemental car-

0097-6156/81/0167-0223$05.00/0

bon is difficult, and, as Grosjean (13) has pointed out, current analytical definitions of elemental carbon are "operational" or method-dependent.

In this paper we report on recent progress achieved in our laboratory with regard to the separate and quantitative measurement of organic and elemental carbon in aerosol samples. Our initial approach (4) was to heat a segment of an aerosol-containing glass or quartz fiber filter to 600°C in a helium atmosphere. Volatilized organic compounds were oxidized to CO_2 in an oxidation oven, reduced to CH_4, and measured by a flame ionization detector (FID). Residual carbon, which was assumed to be elemental carbon, was oxidized to CO_2 by the addition of O_2 to the combustion oven and measured as above. It was discovered, however, that if the filter was removed from the oven after the organic analysis step-- but before the addition of O_2, it was noticeably darker than at the beginning of the analysis. This was indicative of the pyrolytic formation of elemental carbon during the organic analysis (5).

Since that discovery a major aspect of our research effort has been to develop a method to account for this unwanted pyrolytic conversion of organic to elemental carbon. This report describes a combined thermal-optical instrument in which the reflectance of the filter sample is continuously monitored during the thermal analysis. Dod et al. (14) have also reported a combustion technique combined with optical transmission.

Experimental

The carbon analyzer consists of three principal parts: the combustion system, the laser reflectance system, and the microprocessor control. The combustion system is shown in Figures 1 and 2. Four filter disks, each 0.25 cm^2 in area, are mounted vertically in a quartz boat which is located in the loading section of the combustion oven. The oven is purged with a 2% O_2-98% He mixture, and the temperature of the heating zone is set to 350°C. The boat is then inserted into the heating zone in which oxidation and volatilization of organic carbon into the flowing O_2-He stream occur. The volatilized organic carbon is transported through the oxidation zone, which is a bed of granular MnO_2 at 950°C. This results in the oxidation of the organic carbon to CO_2 which is subsequently reduced to CH_4 in the Ni/firebrick (450°C) methanator and measured by a flame ionization detector.

The carrier gas is then changed to He, and after purging to remove O_2, the heating zone temperature is raised to 600°C. This produces a further volatilization of organic carbon which is measured as above. The purpose of the two-step organic analysis is to minimize the problem of pyrolytic conversion of organic to elemental carbon. As discussed below, however, it has not been possible to eliminate this completely, and a significant amount of pyrolytic conversion occurs during the 600°C volatilization.

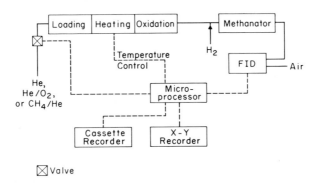

Figure 1.   Block diagram of C analyzer.   The valve symbol represents a complex network of valves and plumbing.

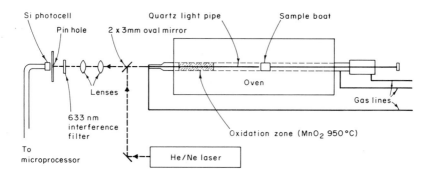

Figure 2.   Laser reflectance system.   The light pipe is a 3-mm-diameter quartz rod.

The remaining carbon on the filter includes elemental carbon originally on the filter, elemental carbon produced pyrolytically during the organic analysis, and also possibly other forms of carbon which have high temperature stability. This carbon is measured by lowering the heating zone temperature to 400°C and changing the carrier gas to the O_2-He mixture. Oxidation of the carbon to CO_2 is carried out sequentially at 400, 500, and 600°C. This three-step process results in a relatively slow oxidation of elemental carbon and is used in the correction for pyrolytic conversion as described below. When the elemental carbon measurement is complete, the temperature of the heating zone is lowered to 350°C in preparation for the next sample. As the cooling is taking place, a known amount of CH_4 is injected into the oven. The resultant FID response provides a calibration for that run.

The correction for the pyrolytic production of elemental carbon is achieved with the laser reflectance system shown in Figure 2. Light (633 nm) from a He-Ne laser is reflected off the 2x3mm oval mirror and down the quartz light pipe to the filter sample which is mounted vertically with the aerosol deposit facing the light pipe. Much of the returning, diffusely reflected light misses the oval mirror and is collected by the lens system and measured by the photocell. Light which is specularly reflected by the ends of the light pipe is deflected away from the photocell by the oval mirror. Thus the photocell sees primarily light which has been reflected from the filter.

The correction for the pyrolytic production of elemental carbon is accomplished by measuring the amount of elemental carbon oxidation necessary to return the filter reflectance to its initial value. This is facilitated by the three-step elemental carbon oxidation which produces a relatively slow initial rise in the reflectance. A typical output is shown in Figure 3. The pyrolysis correction corresponds to the shaded area which is added to peaks 1 and 2 to give the corrected value for organic carbon. This procedure assumes that the mass absorption coefficient of the pyrolytically produced elemental carbon is the same as that of the original elemental carbon. Research to test this assumption is continuing.

The reflectance system also provides a test of the effectiveness of purging prior to the 600°C/He volatilization of organic carbon. Residual O_2 at 600°C will oxidize elemental carbon and produce an increase in the reflectance of the filter. In the event that such an increase is observed, the sample would be rerun, and if this behaviour persists, the system would be checked for leaks or other malfunctions. Thus, the reflectance system plays an important quality assurance role in the analysis.

The complete analytical system is under the control of a Motorola 6802 microprocessor. All switching of gas flows, timing, temperature control, error detection, analog to digital conversion, FID current measurement, signal integration and manipulation, and data storage and transfer are controlled by this system.

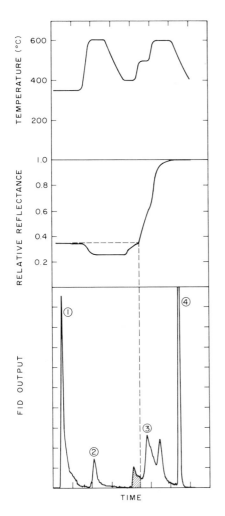

Figure 3.   Analytical response.   Peaks 1, 2, and the shaded portion of 3 correspond to organic C; the unshaded portion of Peak 3 is elemental C.   The shaded portion of Peak 3 constitutes the correction for pyrolytic conversion of organic to elemental C.   Peak 4 is the calibration peak.

A block diagram of the microcomputer system is shown in Figure 4.
Details of this system are discussed by Johnson (15).
Carbonate carbon is determined on a separate filter segment
by measurement of the CO_2 evolved upon acidification with 20 μl
of 1% H_3PO_4. Because carbonates also respond in the organic mode
during thermal analysis, the organic carbon concentration must be
corrected by subtracting the carbonate carbon concentration.

Instrument Validation

Both the accuracy and precision of analysis were investigated
by a number of experiments. Known amounts of pure compounds were
added to quartz fiber filter segments which were analyzed in the
usual manner. The compounds included tetracosane, tetratriacon-
tane, coronene, perylene, glutaric acid, oleic acid, stearic acid,
dioctylphthalate, and mannitol. The average percent recovery was
99±6%, and only mannitol, a sugar, showed a significant amount
(6%) of conversion of organic to elemental carbon during the organic analysis. Analysis of lampblack gave 1% as organic carbon
and 98% as elemental carbon for a total recovery of 99%.
More complex substances, however, showed higher degrees of
pyrolytic conversion. In aerosol samples collected from the combustion of distillate and residual oil, 31 and 25% respectively
of the organic carbon underwent pyrolytic conversion to elemental
carbon. Lesser amounts of conversion were observed for leaded and
unleaded auto exhaust samples, and no conversion occurred in a
diesel truck exhaust sample. Biological samples also showed large
degrees of conversion; e.g., 45% of the carbon associated with
wood fiber was pyrolytically converted to elemental carbon.
High degrees of conversion were also observed in ambient samples. In approximately 200 filters from 9 urban sites an average
of 22% of the organic carbon was pyrolytically converted to elemental carbon. As a fraction of elemental carbon this corresponded to 23%. Thus, the correction for pyrolytic conversion is significant and cannot be neglected.
To elucidate the nature of the carbon which is pyrolytically
converted to elemental carbon, solvent extraction studies were
performed. A filter segment was placed in a stainless steel filter holder, and the solvent was forced through the filter by a sy-
ringe pump at the rate of 0.5 cm^3/min for 60 minutes. This pro-
cedure--rather than more conventional ones (e.g., Soxhlet extraction)-- was used to minimize the wash-off of insoluble particles
from the filter. Both extracted and unextracted filter segments
were subsequently analyzed in the carbon analyzer. Although the
results varied for different solvents and different filters, the
principal finding was that up to 80% of the pyrolytic conversion
could be eliminated by organic solvent extraction. (Similar removal efficiencies were found for total organic carbon.) These
results confirm the organic origin of the pyrolytically produced
elemental carbon. Solvent extraction studies to verify the speciation into organic and elemental carbon are continuing.

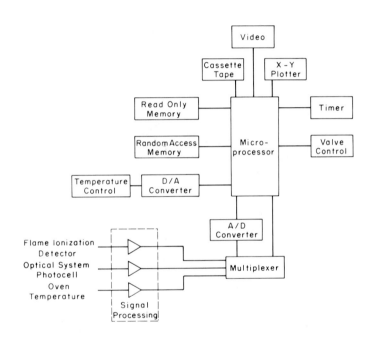

Figure 4. Block diagram of microprocessor system

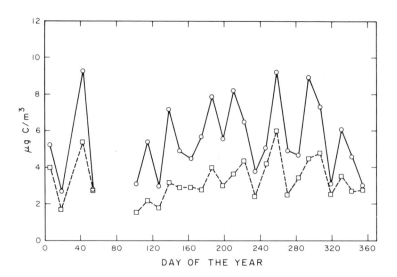

*Figure 5.   Organic (–○–) and elemental (– –□– –) C concentrations in Detroit,
MI—1975.   The value r is the correlation coefficient between the organic and elemental C concentrations (r = 0.83).*

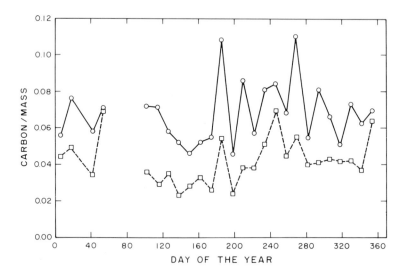

*Figure 6.   Mass fraction of organic (–○–) and elemental (– –□– –) C in Detroit,
MI—aerosol in 1975.   Mass is the total mass concentration of aerosol as measured
by a high-volume sampler; r is the correlation coefficient between organic and elemental mass fractions (r = 0.60).*

The precision of the pyrolytic conversion correction has been assessed by repeated analysis of one high volume filter. For a filter with 36 $\mu g/cm^2$ of organic carbon and 25 $\mu g/cm^2$ of elemental carbon one standard deviation corresponded to ±10% in both the organic and elemental modes.

The analytical sensitivity is limited by uncertainties in the response to blank filters. For glass fiber filters (Gelman A/E) the blank values are 2.8±1.4 $\mu gC/cm^2$ for organic carbon and 0.2 $\mu gC/cm^2$ for elemental carbon. For Pallflex QAST the respective values are 1.0±0.5 and 0.3±0.2 $\mu gC/cm^2$.

Results on Atmospheric Samples

The carbon analyzer has been used to analyze filters from 42 urban sites and 22 non-urban sites in the United States. These filters were obtained from the National Air Surveillance Network (NASN) filter bank for 1975. Carbon concentrations and mass fractions for Detroit, Michigan, are shown in Figures 5 and 6. Both the organic and elemental carbon concentrations are highly variable, and no seasonal trends are apparent. For this site elemental carbon constituted 38% of total aerosol carbon. Typical values for other sites ranged between 35 and 55%.

Filters have also been analyzed from sites in the vicinity of a Lurgi coal gasifier (16) and in the Ohio River Valley. Both these results and the NASN results will be discussed in detail elsewhere.

Summary

An instrument employing both thermal and optical measurements has been developed for the analysis of organic and elemental carbon on glass or quartz fiber filters. During the thermal analysis the reflectance of the filter is continuously monitored to permit a quantitative correction for the pyrolytic production of elemental carbon which occurs during the organic analysis. This correction has been shown to be significant--typically 22% of both organic and elemental carbon. The instrument is completely automated and is under the control of a microprocessor system. It has been used to measure organic and elemental carbon concentrations from both urban and rural sites in the United States. For urban sites the concentrations of both forms of carbon are highly variable with organic carbon slightly more abundant.

Acknowledgment

This work was supported in part by National Science Foundation Grant No. PFR-7824554 and U. S. Environmental Protection Agency Grant No. R806274.

Literature Cited

1. Appel, B.R.; Colodny, P.; Wesolowski, J.J. "Analysis of carbonaceous materials in Southern California aerosols". Environ.
Sci. Technol., 1976, 10, 359-363.
2. Appel, B.R., Hoffer, E.M.; Kothny, E.L., Wall, S.M.; Haik, M.;
Knights, R.L. "Analysis of carbonaceous material in Southern
California atmospheric aerosols 2". Environ. Sci. Technol., 1979,
13, 98-104.
3. Pierson, W.R.; Russell, P.A. "Aerosol carbon in the Denver
area in November 1973". Atmos. Environ., 1979, 13, 1623-1628.
4. Johnson, R.L.; Huntzicker, J.J. "Analysis of volatilizable
and elemental carbon in ambient aerosols", in "Proceedings: Carbonaceous Particles in the Atmosphere", T. Novakov, Ed., Lawrence Berkeley Laboratory, Berkeley, California, June 1979, 10-13.
5. Johnson, R.L.; Shah, J.J., Huntzicker, J.J. "Analysis of organic, elemental, and carbonate carbon in ambient aerosols", in
"Sampling and Analysis of Toxic Organics in the Atmosphere", American Society for Testing and Materials, STP 721, Philadelphia,
PA, 1980, pp. 111-119.
6. Cadle, S.H.; Groblicki, J.P.; Stroup, D.P. "An automated carbon analyzer for particulate samples", presented at the Second
Chemical Congress of the North American Continent, Las Vegas, Nevada., August 1980.
7. Grosjean, D.; Heisler, S.; Fung, K.; Mueller, P.; Hidy, G.
"Particulate organic carbon in urban air: concentrations, size
distribution and temporal variations", presented at the American
Institute of Chemical Engineers 72nd Annual Meeting, San Francisco, California, November 1979.
8. Daisey, J.M.; Leyko, M.A.; Kleinman, M.T.; Hoffman, E. "The
nature of the organic fraction of the New York City Summer Aerosol", Ann. New York Acad. Sci., 1979, 322, 125-141.
9. McCarthy, R.; Moore, C.E. "Determination of free carbon in
atmospheric dust", Anal. Chem., 1952, 24, 411-412.
10. Kukreja, V.P.; Bove, J.L. "Determination of free carbon collected on high volume glass fiber filter", Environ. Sci. Technol.,
1976, 10, 187-189.
11. Pimenta, J.A.; Wood, G.R. "Determination of free and total
carbon in suspended air particulate matter collected on glass fiber filters", Environ. Sci. Technol., 1980, 14, 556-561.
12. Delumyea, R.G.; Chu, L.-C.; Macias, E.S. "Determination of
elemental carbon component of soot in ambient aerosol samples",
Atmos. Environ., 1980, 14, 647-652.
13. Grosjean, D., comments at the Second Chemical Congress of the
North American Continent, Las Vegas, Nevada, August 1980.
14. Dod, R.L.; Rosen, H.; Novakov, T. "Optico-thermal analysis
of the carbonaceous fraction of aerosol particles", in "Atmospheric Aerosol Research: Annual Report 1977-78", Lawrence Berkeley Laboratory, Berkeley, California LBL-8696, pp.2-10.
15. Johnson, R.L. "Development and evaluation of a thermal-opti-

cal method for the analysis of carbonaceous aerosols", M.S. thesis, Oregon Graduate Center, 1981.
16. Huntzicker, J.J.; Johnson, R.L.; Shah, J.J. "Carbonaceous aerosol in the vicinity of a Lurgi gasifier", presented at the Second Chemical Congress of the North American Continent, Las Vegas, Nevada, August 1980.

RECEIVED March 10, 1981.

Wintertime Carbonaceous Aerosols in Los Angeles

An Exploration of the Role of Elemental Carbon

M. H. CONKLIN and G. R. CASS

California Institute of Technology, Pasadena, CA 91125

L.-C. CHU and E. S. MACIAS

Washington University, St. Louis, MO 63130

Aerosol carbon concentrations have been measured under wintertime conditions at two sites in the Los Angeles basin. Samples were analyzed for total carbon by the Gamma Ray Analysis of Light Elements (GRALE) technique and for elemental carbon by reflectance. Total carbon concentrations at downtown Los Angeles averaged 24 μgm^{-3} and elemental carbon concentrations averaged 9 μgm^{-3} during the 7 am to 1 pm period. The light absorption efficiency of ambient elemental carbon particles was determined to be 11.9 $\pm$ 0.9 m^2g^{-1}, and it was estimated that light absorption by elemental carbon could account for up to 17% of total light extinction in downtown Los Angeles. Baseline conditions for elemental carbon concentrations in the Los Angeles area prior to the introduction of large numbers of diesel passenger cars into the vehicle fleet are established by these measurements. A method for reconstruction of a long time series of past historical air monitoring data on elemental carbon concentrations is described.

Primary carbon particles consisting of volatile organics plus soot are emitted from most combustion processes. The soot fraction of these pollutant emissions consists principally of graphitic or elemental carbon and adsorbed polynuclear aromatic hydrocarbons (1). Graphitic carbon particles are thought to be the predominant light absorbing aerosol species in the atmosphere (2), and have been related to visibility deterioration in the Denver region (3). Carbonaceous aerosols are observed to accumulate in the respirable submicron particle size range (4,5). Soots have been shown to be carcinogenic in experimental animal studies and hence are of considerable public health concern (6).

0097-6156/81/0167-0235$05.00/0

There are presently no large historical air quality monitoring data bases for elemental carbon. The development of an historical data base is necessary for the design of emission control
strategies because it is important to understand the nature of an
existing air quality problem before the effect of altered emissions can be evaluated. Numerous short-term studies completed in
St. Louis (7), Los Angeles (8,9), Berkeley (10), Denver (3), and
Austria (11) show that elemental carbon aerosols are present in
significant quantities, but data sufficient to define long-term
averages and to evaluate spatial and temporal trends over periods
of years are absent.

The lack of a complete data base on soot concentrations in
Los Angeles is of particular importance. As noted by Pierson
(12), Los Angeles air quality is likely to be heavily affected by
increased soot emissions if large numbers of diesel passenger cars
are introduced into the vehicle fleet. It is important to characterize existing air quality before this occurs.

The purpose of this paper is to report progress to date on
the reconstruction of a long-term historical data base for elemental carbon concentrations in the Los Angeles basin. Since the
mid-1950's, the Los Angeles Air Pollution Control District and its
successor agencies have collected consecutive hourly average data
using instruments which measure particulate matter in Km units.
The Km unit is a reflectance measurement of the "blackness" of
particulate matter deposited on a filter. The darkened appearance
of ambient particulate samples has been attributed to the elemental carbon content of the sample (2). A working hypothesis at the
outset of this project was that the Km sampler data can be related
to elemental carbon concentrations.

To verify this hypothesis, low volume samplers have been
operated in parallel with the air pollution control district's Km
samplers. The samples collected were analyzed optically for elemental carbon and by the Gamma Ray Analysis of Light Elements
(GRALE) technique for total carbon. These data were used to
assess the concentration of elemental and total carbon aerosols
present during the winter months in Los Angeles. It was established that the Km samplers can be calibrated to read elemental
carbon concentrations. This calibration can be used to reconstruct historical elemental carbon levels at seven sites in Los
Angeles.

Experimental Design

A sampling schedule was designed which would encounter peak
levels of ambient elemental carbon. Peak levels were of interest
for two reasons: to insure that filter loadings would be within
the appropriate range for maximum analytical sensitivity and to
expand the range of values over which the Km samplers could be
calibrated. To determine when peak elemental carbon concentrations would be likely to occur, historical data on carbon monoxide

and total oxides of nitrogen concentrations in Los Angeles were examined as tracers for the presence of primary combustion pollutants. It was found that concentration maxima for these pollutants occur during the morning peak traffic hours throughout the winter months. Using this information, the sampling schedule was set for 10 weekday mornings during the period January 15 to February 26, 1980.

Low-volume samplers were operated in parallel with South Coast Air Quality Management District air monitoring equipment at the District's Pasadena and downtown Los Angeles stations. As shown in Figure 1, the downtown Los Angeles sampling site presently is located within a ring of freeway interchanges. It is in an industrial area to the northeast of the central business district. Railroad switching yards are east of that site. The Pasadena monitoring station is situated in a mixed residential/commercial district two blocks south of the Interstate 210 Freeway.

Three pump and filter assemblies were operated simultaneously at each location. Sample duration of 51.5 minutes each hour and sample inlet line configuration replicated the Km samplers. Sampling began at 700 hours Pacific Standard Time (PST) and continued over consecutive hours until 1300 PST.

Filter substrates were chosen that were suited for subsequent determinations of aerosol mass, total carbon, elemental carbon, trace metals and aerosol light absorption coefficient (b_{abs}). The three filter assemblies contained, a 47 mm Pallflex Tissuquartz filter (2500 QAO), a 47 mm Nuclepore polycarbonate filter (0.40 µm pore size) and a 13 mm Pallflex Tissuquartz filter (2500 QAO).

The 47 mm Pallflex Tissuquartz filters were chosen because of their low carbon content (13). Filter preparation consisted of prefiring these filters to 900°C in air for 1-1/2 hours to further reduce the carbon levels. These filters were masked leaving a 0.32 cm^2 area exposed to the air flow. Ambient air was drawn through each filter at a rate of approximately 12 l/min. Following sample collection each of the filters was placed in a plastic petri dish, sealed and chilled. Filter deposits were subsequently analyzed for total carbon by the Gamma Ray Analysis of Light Elements (GRALE) technique (13). Elemental carbon determination on the same filters was obtained by reflectance calibrated against the carbon content of butane soot standards which had been heated gradually in air to 300° C and held at that temperature for greater than 10 min in order to remove organics present. The elemental carbon concentration of ambient samples also can be obtained by the GRALE technique after similar heat treatment to eliminate volatile organics. The equivalence of the reflectance and GRALE techniques for measurement of ambient elemental carbon concentrations is demonstrated by Delumyea, Chu and Macias (7). Their experiments in St. Louis show that the reflectance method calibrated against heated butane soot will reproduce the elemental carbon concentration (EC) of ambient samples as measured by GRALE with a near unit slope:

$$EC_{GRALE} = (0.9 \pm 0.2) \ EC_{REFLECT} + (3 \pm 4) \tag{1}$$

and correlation coefficient of 0.96. They found that the butane soot standards have reflectance properties representative of ambient elemental carbon but that a small amount of non-elemental carbon may remain in samples and standards after heat treatment in air at 300° C.

The Nuclepore filters used are translucent, light in weight and non-hygroscopic, which makes them desirable for mass measurements and light absorption measurements. Filter preparation consisted of removing any static charge and weighing each filter. Following sample collection at a flowrate of approximately 20 l/min, the filters were sealed and chilled. After reweighing to determine the collected aerosol mass, the Nuclepore polycarbonate filters were divided into pie shaped wedges for determination of b_{abs} and trace metal concentrations. Measurement of b_{abs} was performed using the opal glass integrating plate technique (14) as modified by Ouimette (15). In this procedure an exposed filter is illuminated by coherent light at a wavelength of 0.6328 μm transmitted through a 0.25 mm diameter optical fiber. Light transmitted through the sample is detected by a photo transistor. Light absorption by the aerosol deposit is calculated from the ratio of the intensity of light transmitted through the aerosol deposit to that transmitted through a blank edge of the same filter:

$$b_{abs} = \frac{A}{V} \ ln \ \frac{I_o}{I} \tag{2}$$

where

b_{abs} is the aerosol light absorption coefficient (m^{-1});

A   is the filter surface area covered by the aerosol deposit (m^2);

V   is the volume of air drawn through the filter (m^3);

I_o is the intensity of light transmitted through an unexposed portion of the filter; and

I   is the intensity of light transmitted through the aerosol deposit on the exposed area of the filter.

At least five replicate measurements of transmission were made on different portions of the filter sample to account for any variations in the filter loading. The standard deviation of the transmission measurements made on a single sample ranged between 1% and 4% of the mean value for that sample. The third sample taken each hour was collected on a 13 mm Pallflex Tissuquartz (2500 QAO) filter at a flowrate of 20 l/min (nominal). Those filters also were prefired to 900°C for 1-1/2 hours prior to use to lower their carbon content. The 13 mm quartz filters were archived for future use.

Hourly measurements of other monitored pollutants and meteorological conditions were obtained from instruments located at each sampling station:

NO_x, NO, NO_2 (Thermoelectron Corp Model 14 B/E)

CO (Mine Safety Appliances Co LIRA (Pasadena) and Bendix Model 8501-SCA (Los Angeles))

Particulate matter in Km units (A.R. Chaney Chemical Laboratories, Inc., Autosampler)

Wind Speed

Wind Direction

Wintertime Carbonaceous Aerosols in Los Angeles

The average total carbon and elemental carbon concentrations observed at downtown Los Angeles and Pasadena are illustrated in Figure 2. Values shown are averages of one-hour samples taken during the period 7 am to 1 pm. Both elemental and total carbon concentrations are higher in downtown Los Angeles. This was expected as elemental carbon aerosol is a primary pollutant emitted from motor vehicles and stationary combustion sources. Figure 3 shows the diurnal trend in elemental carbon concentrations measured at Pasadena and Los Angeles. Elemental carbon concentrations decline from early morning until mid-day in a manner similar to hourly traffic density (Figure 4).

Wintertime samples at Pasadena and downtown Los Angeles from the present study show elemental carbon to average about 31% and 37% of total carbon, respectively. Considerable scatter in the ratio of elemental to total carbon was observed between individual short-term samples. These data can be compared to winter studies in other cities. Samples taken in Denver during November 1973 showed total carbon concentrations averaging 15.7 μgm^{-3}, of which about 68% could not be removed by solvent extraction. The insoluble carbon present was considered as an upper limit on elemental carbon (3). Gundel (10) reported insoluble carbon concentrations averaging 48% of total carbon present in Berkeley, California during the winter of 1978. Each of these studies would confirm that elemental carbon forms a substantial fraction of the carbon containing particulate matter observed in cities during the winter months.

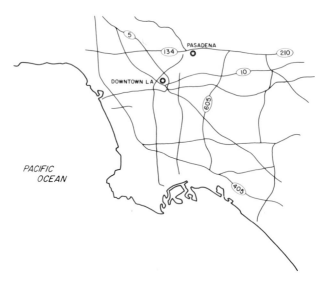

Figure 1.   The metropolitan Los Angeles area showing sampling sites in relation to
the local freeway network ((✪) air monitoring station)

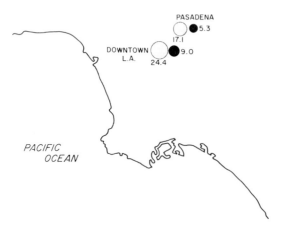

Figure 2.   Average total ((O) by grale) and elemental ((●) by reflectance) C concentrations observed at downtown Los Angeles and Pasadena during the hours
7 am to 1 pm, January 15–February 26, 1980 (units in μg/m³)

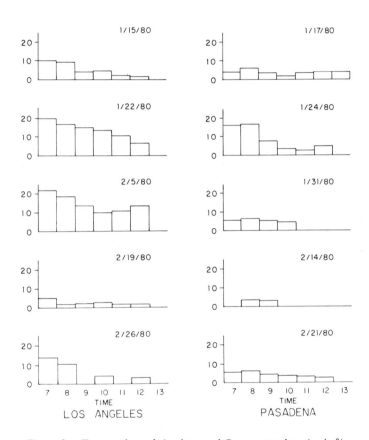

Figure 3. Temporal trends in elemental C concentrations ($\mu g/m^3$)

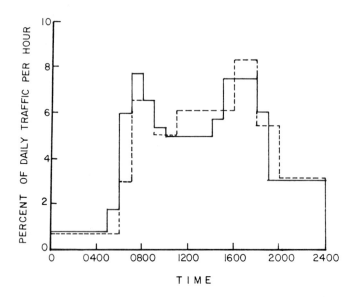

Figure 4.   Diurnal variation of Los Angeles traffic flow, after Nordsieck (23)
((——) freeways; (– – –) non-freeways)

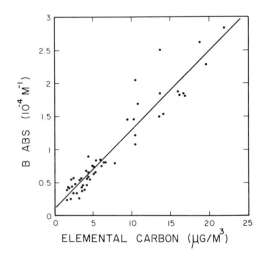

Figure 5.   Aerosol light absorption as a function of elemental C concentration

It might be expected that the ratio of elemental to total carbon would be higher in the winter than during the summer in Los Angeles due to increased production of secondary aerosols during the summer season. Appel et al. (8) studied the composition of Los Angeles area carbonaceous aerosols during the summer and fall of 1973. An upper limit estimate of elemental carbon present was obtained by analysis of carbon species which were insoluble during solvent extraction. Estimates of elemental carbon present ranged from 0% to 41% of total carbon, with a mean value of 22.3%. A later study during July 1975 at Pasadena, Pomona and Riverside showed unextractable carbon ranging from 51.5% to 41.4% of total carbon, with a mean value of 43.5% (9). As noted by Gundel (10), Appel et al.´s (9) summer results do not differ greatly from the Berkeley aerosol in the winter nor do they differ greatly from our winter results. A precautionary note should be added because equating insoluble carbonaceous material with elemental carbon provides only an upper limit to actual elemental carbon present (16).

Graphitic carbon particles are thought to be the most abundant light-absorbing aerosol species in the atmosphere (2). Comparison of elemental carbon concentrations from the present study to measurements of b_{abs} is presented in Figure 5. A regression line drawn through those data has the equation:

$$b_{abs} = 0.119 \text{ EC} + 0.10 \qquad n = 53 \qquad (3)$$
$$(0.005) \qquad (0.047) \qquad \rho = 0.95$$

where

b_{abs} is the aerosol absorption coefficient $(10^{-4} m^{-1})$;

and

EC is elemental carbon in (μgm^{-3}).

Values in parentheses represent one standard error of those coefficient estimates.

An upper limit to the absorption efficiency of the elemental carbon particles sampled can be obtained if it is assumed that all aerosol light absorption is due to elemental carbon. In that case the specific absorption from the slope of the line in Figure 5 is about $11.9 \pm 0.9 \ m^2 g^{-1}$. That value is higher than would be expected for absorption by single elemental carbon spheres (17) but is within the range of values which could be calculated theoretically for agglomerates of fine carbon particles (18). Roessler and Faxvog (18) have observed a specific absorption for acetylene soot of $8.3 \pm 0.9 \ m^2 g^{-1}$. Macias et al. (7) noted that butane soot contains an appreciable amount of organic carbon, about 25%. If acetylene soot were similarly composed, then Roessler and Faxvog´s (18) specific absorption for acetylene soot would be higher if based on elemental carbon mass alone.

The long-term average total extinction coefficient at down-
town Los Angeles is $6.62 \ 10^{-4} \ m^{-1}$ ($\underline{19}$) as inferred from daytime
measurements of prevailing visibility. Elemental carbon present
at about 9 μgm^{-3} with a specific absorption of 11.9 $m^2 \ g^{-1}$ would
account for about 17% of total light extinction at downtown Los
Angeles.

Elemental carbon concentrations are the result of incomplete
combustion. These primary carbon particles should closely track
the gas phase products of combustion processes. Figures 6 and 7
show that this is indeed the case. Figure 6 presents the relationship between 1-hour average samples of elemental carbon and
total oxides of nitrogen. Figure 7 shows the relationship between
elemental carbon and CO. One reason for the scatter in the CO
results is that the data are only reported to the nearest ppm.

Estimation of an Historical Data Base for Elemental Carbon Concentrations

Nearly 23 years of hourly observations on particulate matter
concentrations have been collected at downtown Los Angeles and at
six other sites in the Los Angeles area using a tape sampler calibrated to read in Km units. The design of these samplers is as
described by Hall ($\underline{20}$). Ambient air is drawn through a one-
square-inch (5.07 cm^2) area of filter paper at a rate of 25 ft^3
per hour (11.8 lpm) for 51.5 min of each hour. The darkness of
the spot developed on the filter is measured by the ratio of the
intensity of the reflected light from clean white filter paper,
R_o, to reflected light from the aerosol deposit, R. As used by
the air pollution control district, the Km unit is related to
reflectance by the formula ($\underline{21}$).

$$Km = \frac{2280}{v_m} \ log_{10} \ \frac{R_o}{R} \qquad (4)$$

where v_m is the measured volume of air sampled during each 51.5
min period (std. cu. ft., at 1 atm, $70°F$).

If the blackness of the particulate matter collected on a
filter is due to the graphitic carbon content of the sample, then
the Km unit should convert to ambient elemental carbon concentrations. The form of that translation is apparent from the definition of the Km unit. Elemental carbon concentration measurements
made by laboratory reflectometers calibrated against heated butane
soot standards show that elemental carbon concentrations are
linearly related to the log of the reflectance ratio R_o/R. Aerosol loadings stated in Km units should be directly proportional to
elemental carbon concentrations sampled.

The series of field experiments described in this paper was
conducted with the principal objective of obtaining the data
needed to calibrate the Km sampler records in terms of elemental
carbon concentrations. The results of the sampling effort are
shown in Figure 8. Elemental carbon concentrations and Km sampler

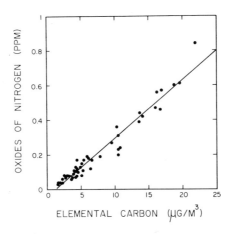

Figure 6. Relationship between oxides of N and elemental C concentrations

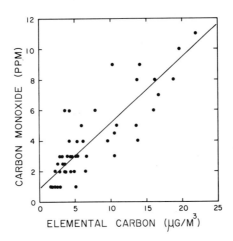

Figure 7. Relationship between CO measurements and elemental C concentrations

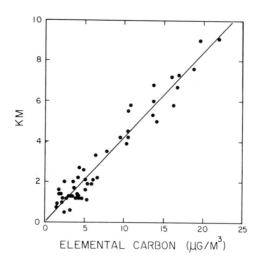

Figure 8.   Calibration curve: Km units as a function of elemental C concentration

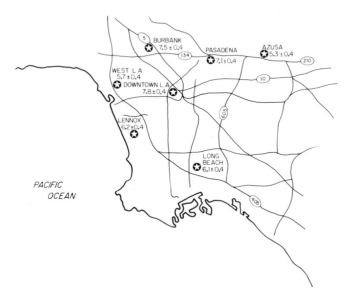

Figure 9.   Average elemental C concentrations by the reflectance method in the Los Angeles Basin 1958–1972, in μgm^{-3} ((☼) air monitoring station).   Error bounds represent a 95% confidence interval on the long-term means.

response are linearly related with a correlation coefficient of 0.97. A regression line drawn through those data has the equation

$$EC = 2.24 \ Km + 0.38 \qquad n = 53 \qquad (5)$$
$$(0.078) \qquad (0.30) \qquad \rho = 97$$

where EC is elemental carbon (μgm^{-3}). Values in parentheses represent one standard error on those coefficient estimates.

Using Equation (5) elemental carbon concentrations can be estimated from historical Km sampler data at seven sites in the Los Angeles area. Reprocessing of that data forms a next step in this project. Some insight can be gained from existing reports. Km data taken during the period 1958-1972 have been displayed by Phadke et al. (22). Long-term mean Km values for this period were reported for each of the sampling stations. When converted to elemental carbon concentrations, those data are shown in Figure 9. Long-term mean elemental carbon concentrations in the Los Angeles basin if measured by the reflectance method would have been reported to range from 5.3 $\pm$ 0.4 to 7.8 $\pm$ 0.4 μgm^{-3} during the period shown, with the highest concentrations observed in downtown Los Angeles. Uncertainties arising from the measurement of elemental carbon by reflectance were noted previously and can be quantified by reference to Delumyea, Chu and Macias (7).

Conclusions

Aerosol carbon concentrations have been measured at two sites in the Los Angeles basin. Samples were analyzed for total carbon content and for elemental carbon content by the Gamma Ray Analysis of Light Elements technique and by several optical methods. Elemental carbon was shown to constitute a substantial fraction of total carbonaceous aerosol mass in the wintertime in Los Angeles.

Samples collected were used to determine the light absorption efficiency (11.9 $\pm$ 0.9 $m^2 g^{-1}$) of ambient elemental carbon particles. This value is slightly higher than the mass absorption efficiency previously reported for soot samples. An explanation for this is that soot is composed in part of organic carbon-containing materials and other weakly-absorbing species, thus lowering the absorption coefficient per unit total mass. Using the absorption efficiency and the elemental carbon concentrations measured in downtown Los Angeles ($9 \mu gm^{-3}$), it was calculated that light absorption by elemental carbon can account for up to 17% of total light extinction observed at that location. This indicates that elemental carbon has a significant role in visibility degradation in Los Angeles.

Air pollution control agencies in the Los Angeles area have accumulated 22 years of consecutive hourly samples of "particulate matter measured in Km units." It was shown that the historical Km data should be proportional to elemental carbon concentrations present, and the Km samplers were calibrated to read elemental

carbon concentrations using parallel samples taken during this
project. From historical Km data, 15-year average values of elemental carbon concentrations were estimated at seven sites in the
Los Angeles area. These data show that the downtown Los Angeles
monitoring station historically has experienced the highest levels
of elemental carbon, with 8 μgm^{-3} as a long-term average.

Acknowledgement

This work has been supported in part by the U.S. Department
of Energy under Grant No. DE-FG03-76AD01305, by the U.S. Environmental Protection Agency under Grant No. R806005, and by the Mel-
lon Foundation. Thanks are due to the staff of the South Coast
Air Quality Management District for their cooperation throughout
the course of this project.

Literature Cited

1. Wagner, H.G.G., Soot Formation in Combustion, "Seventeenth
    Symposium (International) on Combustion;" The Combustion
    Institute: Pittsburgh, Pennsylvania, 1978; pp. 3-19.

2. Rosen, H.; Hansen, A.D.A.; Dod, R.L.; Novakov, T., "Applica-
    tion of of the Optical Absorption Technique to the Characteri-
    zation of the Carbonaceous Component of Ambient and Source
    Particulate Samples;" Lawrence Berkeley Laboratory: Berkeley,
    California, 1977.

3. Pierson, W.R.; Russell, P.A., Aerosol Carbon in the Denver
    Area in November 1973, Atmospheric Environment, 1979, 13,
    1623-1628.

4. Hidy, G.M. et al., "Characterization of Aerosols in California
    (ACHEX), Volume IV;" Rockwell International Science Center:
    Thousand Oaks, California, 1974; p. 3-49.

5. Task Group on Lung Dynamics, Deposition and Retention Models
    for Internal Dosimetry of the Human Respiratory Tract, Health
    Physics, 1966, 12, 173-207.

6. IARC Working Group, An Evaluation of Chemicals and Industrial
    Processes Associated with Cancer in Humans Based on Human and
    Animal Data: IARC Monographs Volumes 1 to 20, Cancer Research,
    1980, 40, 1-12.

7. Delumyea, R.G.; Chu, L-C.; Macias, E.S., Determination of Ele-
    mental Carbon Component of Soot in Ambient Aerosol Samples,
    Atmospheric Environment, 1980, 14, 647-652.

8. Appel, B.R.; Colodny, P.; Wesolowski, J.J., Analysis of Carbonaceous Materials in Southern California Atmospheric Aerosols, Environ. Sci. Technol., 1976, 10, 359-363.

9. Appel, B.R.; Hoffer, E.M.; Kothny, E.L.; Wall, S.M.; Haik, M., Analysis of Carbonaceous Material in Southern California Atmospheric Aerosols 2, Environ. Sci. Technol., 1979, 13, 98-104.

10. Gundel L. in T. Novakov, Ed., "Proceedings, Carbonaceous Particles in the Atmosphere;" Lawrence Berkeley Laboratory: Berkeley, California, 1978; p. 91.

11. Malissa, H. in T. Novakov, Ed., "Proceedings, Carbonaceous Particles in the Atmosphere;" Lawrence Berkeley Laboratory: Berkeley, California, 1978; pp. 3-9.

12. Pierson, W.R. in T. Novakov, Ed., "Proceedings, Carbonaceous Particles in the Atmosphere;" Lawrence Berkeley Laboratory: Berkeley, California, 1978; pp. 221-228.

13. Macias, E.S.; Radcliffe, C.D.; Lewis, C.W.; Sawicki, C.R., Proton Induced γ-ray Analysis of Atmospheric Aerosols for Carbon, Nitrogen and Sulfur Composition, Analytical Chemistry, 1978, 50, 1120-1124.

14. Lin, C.; Baker, M.; Charlson, R.J., Absorption Coefficient of Atmospheric Aerosol: A Method for Measurement, Applied Optics, 1973, 12, 1356-1363.

15. Ouimette, J.R., "Chemical Species Contributions to the Extinction Coefficient;" Ph.D. Thesis, California Institute of Technology: Pasadena, California, 1980; pp. 210-230.

16. Currie, L.A.; Kunen, S.M.; Voorhees, K.J.; Murphy, R.B.; Koch, W.F., in T. Novakov, Ed., "Proceedings, Carbonaceous Particles in the Atmosphere;" Lawrence Berkeley Laboratory: Berkeley, California, 1978; pp. 36-48.

17. Faxvog, F.R.; Roessler, D.M., Carbon Aerosol Visibility vs. Particle Size Distribution, Applied Optics, 1978, 17, 2612-2616.

18. Roessler, D.M.; Faxvog, F.R., Optoacoustic Measurement of Optical Absorption in Acetylene Smoke, J. Opt. Soc. Am., 1980, 69, 1699-1704.

19. Cass, G.R, On the Relationship Between Sulfate Air Quality and Visibility with Examples in Los Angeles, Atmospheric Environment, 1979, 13, 1069-1084.

20. Hall, S.R., Evaluation of Particulate Concentrations with Collecting Apparatus, Analytical Chemistry, 1952, 24, 996-1000.

21. Holland, W.D.; MacPhee, R.D., "Operating and Servicing the Chaney Aerosol Recorder;" Los Angeles Air Pollution Control District: Los Angeles, California, 1958.

22. Phadke, M.A.; Tiao, G.C.; Grupe, M.; Wu, S.C.; Krug, A.; Liu, S.T., "Los Angeles Aerometric Data on Sulphur Dioxide, Particulate Matter and Sulphate 1955-1972;" Department of Statistics, University of Wisconsin: Madison, Wisconsin, 1975.

23. Nordsieck, R.A., "Air Quality Impacts of Electric Cars in Los Angeles, Appendix A, Pollutant Emissions Estimates and Projections for the South Coast Air Basin;" document RM-1905-A General Research Corporation: Santa Barbara, California, 1974.

RECEIVED March 25, 1981.

Carbonaceous Urban Aerosol—Primary or Secondary?

LIH-CHING CHU and EDWARD S. MACIAS

Department of Chemistry and CAPITA, Washington University, St. Louis, MO 63130

The relative contribution of primary and
secondary carbon to urban aerosol is discussed in
this paper. Some data from the ACHEX study in
Los Angeles have been reexamined using new values
for the carbon and lead emissions. Data on total
carbon, elemental carbon and lead in fine particle
samples collected in St. Louis are presented. Lead
and elemental carbon have been shown to be useful
tracers of primary carbonaceous aerosol. It is
concluded that secondary carbon is most likely to
be a significant portion of the urban carbonaceous
aerosol in the summer and in the middle of the
day. Secondary carbon can best be measured with
short time resolution sampling ($\Delta t \lesssim 6h$).

Ambient aerosols, particularly those with diameters less
than 3μm, are a serious pollution problem. Carbonaceous
material is a major component of the fine particle concentration
(1) and has undergone extensive study in the past few years
(2,3) in large part because of the concern that these particles
play an important role in urban haze and community health.
Particulate carbon in the atmosphere exists predominantly
in three forms: elemental carbon (soot) with attached hydrocarbons; organic compounds; and carbonates. Carbonaceous
urban fine particles are composed mainly of elemental and
organic carbon. These particles can be emitted into the
air directly in the particulate state or condense rapidly
after introduction into the atmosphere from an emission
source (primary aerosol). Alternatively, they can be formed
in the atmosphere by chemical reactions involving gaseous
pollutant precursors (secondary aerosol). The rates of
formation of secondary carbonaceous aerosol and the details
of the formation mechanisms are not well understood. However, an even more fundamental controversy exists regarding

0097-6156/81/0167-0251$05.00/0

the relative amounts of primary and secondary carbonaceous
aerosols in urban air (4,5). It is not possible to review
this debate in detail in this brief space because the literature
is rather extensive. Here we merely mention a few of the arguments which are germane to this paper. The understanding of this
problem is crucial for developing effective strategies to control
particulate carbon pollutants because the controls for primary
and secondary aerosols are quite different.

The results from several groups in the Aerosol Characterization Experiment (ACHEX) (6) support the conclusion that in
Lo; Angeles (LA) during periods of severe pollution, much of
the carbonaceous aerosol is secondary in nature, that is, formed
from homogeneous or heterogeneous reactions in the atmosphere.
Gartrell, Heisler and Friedlander (4) calculated an LA aerosol
carbon balance in which the primary carbon was apportioned by
scaling a particulate emission inventory to automobile exhaust.
Then by assuming these particles were well mixed, the contributions to the aerosol were approximated by scaling to the
automotive exhaust contribution determined by a chemical element
balance. This procedure was applied only to sites where there
were no strong sources of local particulate pollution. The
fraction of carbon in the particulate emissions from various
sources was estimated from source test results or, in the case
of industrial emissions, by assuming that the fraction of
carbon was the same as the fraction measured in the total aerosol.
The difference between measured carbon and the primary carbon
determined from the carbon balance was assumed to be due to
secondary formation. It was concluded from this analysis
that on smoggy sampling days in LA the aerosol carbon was
dominated by secondary aerosol conversion.

Appel, Colodny and Weslowski used a combination of solvent
extractions followed by carbon analysis to estimate the amount
of elemental carbon and primary and secondary organic carbon
aerosol in LA during the ACHEX study (7). They found that
secondary organics were dominant at all inland sampling sites
while elemental carbon and primary organics were of greater
importance at the Dominguez Hills site which is near the coast,
oil refineries, power plants and a freeway. The importance of
secondary carbon aerosol was also suggested from the results
of studies of the detailed speciation of the organic aerosol (8).

More recently Cass, Boone and Macias constructed a very
detailed carbon inventory for Metropolitan Los Angeles in order
to estimate the amount of primary elemental and organic carbon
in this urban area (9). Over 50 source types were included in
this emission inventory. A particulate lead emission inventory
was also constructed and used as a tracer for primary automotive
exhaust. They compared the ratio of organic carbon to elemental
carbon and lead from the emission estimates to that measured in
the atmosphere during winter mornings. In that study the
sampling time and location were chosen in order to measure

an aerosol in which the primary emissions were expected to
dominate. They found excellent agreement between the two
approaches. From that study it can be concluded that primary
carbon dominated in LA during the winter morning sampling
times.

Rosen, Hansen, Dod and Novakov found a high correlation
between optical absorptivity and the particulate carbon loading
in 24-h samples from several California cities (5). Elemental
carbon, a primary pollutant which is directly related to the
absorptivity, was found to be a large fraction of the carbonaceous
aerosol. They were able to place a low limit on the amount of
secondary organic aerosol produced in correlation with ozone.

In this paper we present results which reconcile the
widely different results just discussed ranging from a carbon
aerosol dominated by secondary organic material on the one hand
to a carbon aerosol composed largely of primary carbon compounds
on the other. We have employed an approach which uses lead
or elemental carbon as a tracer for primary emissions and
combines several analysis techniques to reexamine the published
ACHEX data. We also present a new data set from St. Louis
which is analyzed in a similar manner to contrast the aerosol
in a midwestern city with that on the California coast.

ACHEX Data

The ACHEX data base (10) from monitoring stations at
Dominguez Hills, Riverside and West Covina in summer and early
fall of 1973 have been reexamined. Of particular interest in
this context are the aerosol samples (no size fractionation) with
collection times ranging from 2 to 6 hours which were analyzed
for total carbon by combustion and for lead by x-ray flourescence.
These data, plotted in Figure 1, do not show any strong correlation.

We have calculated the emission inventory of particulate
carbon and lead in the Los Angeles Metropolitan area for 1973.
This was calculated in the same manner as that described by
Cass, Boone and Macias (9). The carbon to lead ratio determined
from this inventory can be examined for several limiting cases
ranging from the output of an automobile using leaded fuel
(C/Pb~2) to aerosol in the well mixed LA air basin with significant secondary conversion of carbon (C/Pb>7). This approach
assumes, of course, that lead is present only as a primary
pollutant. For highway traffic composed of vehicles and fuel
types in the same proportions as for the entire urban area, the
ratio of aerosol carbon to lead was calculated to be about 4.
This is about a factor of 2 lower than that calculated for 1980
due to the large number of automobiles which used unleaded fuel
and the lower lead content in leaded fuel in 1980 (9). For the
other extreme where primary emissions from all sources become
well mixed, a situation which is possible because of the long

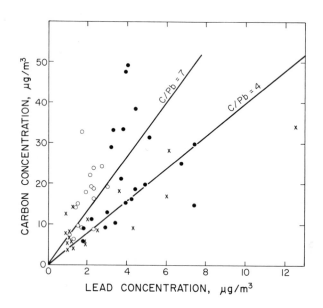

Figure 1. Plots of aerosol C vs. Pb concentrations for Los Angeles in 1973 as determined in the ACHEX study (10). The lines indicate the particulate C/Pb ratio for the well-mixed air basin and the highway signature as explained in the text; (✕) Dominguez Hills; (●) West Covina; (○) Riverside.

retention times in the LA basin, the calculated carbon to lead ratio is about 7. These two limits of primary carbon to lead ratios are also shown as lines in Figure 1. About 40% of the data fall between these two lines and for these data we conclude that no obvious excess carbon due to secondary formation is present. About 20% of the data have C/Pb ratios lower than the highway signature. This may be due to a higher number of vehicles which burn leaded gas, at those sampling times, relative to the basin average (e.g. higher ratio of leaded passenger cars to deisel trucks). Nearly 40% of the points have C/Pb ratios greater than 7. These data clearly have excess carbon relative to lead--almost certainly due to secondary carbon formation. The data with the highest C/Pb ratios are mainly from the inland sites at Riverside and West Covina where photochemical pollution is the most severe.

The effect of secondary carbon formation can be seen even more clearly in Figure 2 where these data are plotted in a slightly different manner. Here the C/Pb ratio is plotted as a function of time of day on three different days. In the early morning the C/Pb ratio approached the highway value at all three sites, indicating dominant primary carbon aerosol at those times. In the middle of the day all sites showed a significant increase in this ratio. Even more to the point, at West Covina the C/Pb values were above 7 from 1000 until 1800 and at Riverside high C/Pb values were measured from 0800 until midnight. These high C/Pb values are indicative of the presence of significant secondary carbon. It should also be noted that if a single 24-h sample had been collected in place of the higher time-resolved samples the effect of secondary carbon would not have been detected in the C/Pb ratio.

Our analysis is in agreement with the earlier conclusions from the ACHEX data that secondary carbon aerosol is significant. One major difference is that our new emission inventory accounts for much more primary carbon than was thought previously, leading to the conclusion that during the early morning primary carbon was dominant, even during very smoggy periods.

Experimental

St. Louis Sample Collection. Ambient aerosols were collected in St. Louis in 6-h intervals with a TWOMASS automated sequential tape sampler. This sampler fractionated the aerosol into two size classes, fine particles having aerodynamic diameters less than 3μm, and coarse particles with diameters greater than 3μm. It was equipped with a beta-attenuation mass monitor to determine fine-particle mass (11). Only the fine particle filter was examined in this study. Pallflex E70 glass-fiber filter tape with a detachable cellulose backing (Pallflex Inc. Putnam, CT) was used with this sampler. An aerosol sampler operating from the same inlet manifold as the

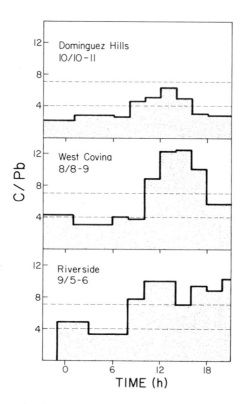

Figure 2. Plot of the C/Pb ratio as a function of time for aerosol samples col-lected at three sampling sites in Los Angeles in 1973. The data are from the ACHEX study (10).

TWOMASS, was used to collect ambient aerosol on Teflon filters (Ghia Corp., Pleasanton, CA). The aerosol inlet used was designed so that large particles did not reach this sampler. Two 6-h samples (0900-1500 and 1500-2100) and one 12-h sample (2100-0900) were collected each day with this sampler.

A total of 154 samples were collected on glass fiber filters during January and February 1978; June and July, 1979; August, 1979; and February and March, 1980. During the last period 24 samples were also collected on teflon filters.

Chemical Analysis. Aerosol carbon, sulfur and nitrogen concentrations were measured on the glass-fiber filters with the gamma ray analysis of light element (GRALE) technique (12). Samples were irradiated with 7-MeV protons in the Washington University 135-cm cyclotron and the γ-rays emitted following inelastic scattering were measured inbeam. The elemental carbon (soot) concentration was measured on the glass-fiber filters by light reflectance from the filter (13). This technique takes advantage of the fact that elemental carbon is the predominant species which is black in the fine particle aerosol. Trace elements were analyzed on teflon filters by particle-induced x-ray emission (PIXE) at the University of California-Davis (14). Of particular interest in this study were the concentrations of lead and bromine.

Results

Fine Aerosol Mass Balance. The temporal variations of fine particle mass, total carbon, elemental carbon, nitrogen, sulfur and b_{scat} for a typical summer period (August 4-10, 1979) and a typical winter period (March 1-9, 1980) are shown in Figures 3 and 4, respectively. The strong correlation between sulfur and nitrogen is seen quite strikingly in these figures and is shown more quantitatively in the scatter plot (Figure 5) of aerosol sulfur and nitrogen concentrations for all samples. These data have a high correlation ($r = 0.91$); the linear least squares fit to the data yields a slope of 1.04 ± 0.04 and inter-cept of $0.5\pm0.2\mu g\ m^{-3}$. This slope is close to the stoichio-metric ratio for ammonium sulfate ($S/N = 1.14$). This correlation and the finding that ambient aerosol sulfur in the eastern U.S. is predominantly in the form of sulfate (1,15), strongly suggest that aerosol sulfur and nitrogen were mainly in the form of ammonium sulfate during this sampling period. However, it is possible that the sulfate was unintentionally neutralized sometime after sample collection. It should be noted that there are several samples with sulfur concentrations significant-ly greater than nitrogen concentrations. This may be due to the presence of an acidic sulfate aerosol or some other sulfur species at those times (16).

A mass balance for fine particle aerosol can be calculated from the carbon, sulfur, nitrogen, and mass concentration data

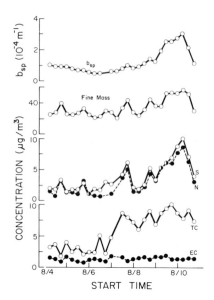

Figure 3. Temporal variations of fine particle mass, total C, elemental C, N, S, and b_{scat} in St. Louis for summer 1979

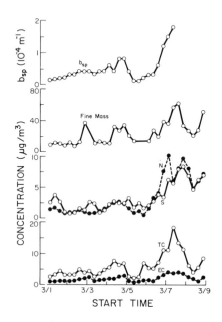

Figure 4. Temporal variations of fine particle mass, total C, elemental C, N, S, and b_{scat} in St. Louis for winter 1980

with the assumptions that aerosol sulfur is in the form of sulfate, that nitrogen is in the form of ammonium ion, and that carbon exists either in the elemental form as soot or in organic compounds. The organic carbon is estimated from the difference between measurements of total carbon and elemental carbon multiplied by 1.5 (e.g. $-CH_2O_{1/4}-$). The fine particle mass balances calculated in this way for summer and winter sampling periods are shown in Figure 6. The average mass concentration was about the same in each sampling period ($\sim$39μg m^{-3}) but several of the measured constituents, particularly elemental carbon, did not remain constant. This increase of elemental carbon in the winter, which may be due to increased space heating and wood burning, somewhat confounds our attempts to use elemental carbon as a tracer for all primary carbonaceous pollutants. However, we have evidence that leads us to think that these data can be used to determine the presence of secondary organic aerosol. For example, the ratio of organic carbon (OC) to elemental carbon (EC) is much greater in the summer sampling period [OC/EC (summer) = 4.3, OC/EC (winter) = 1.8]. We contend that this increase is most likely due to an increase in secondary organic aerosol.

We have also looked for the presence of increased secondary organic aerosol by calculating the fine aerosol mass balance in both summer and winter during periods of high and low sulfate concentrations. Formation of secondary sulfate aerosol can cause elevated levels of sulfates and has been linked to periods of regional scale haziness in the eastern U.S. (17). Our rationale for this analysis procedure is that during periods of increased concentration of secondary sulfate aerosol there might also be an increase in secondary carbon aerosol. The frequency distribution for all samples is shown in Figure 7a. The median sulfate concentration (8.4μg m^{-3}) was chosen to separate the high and low sulfate concentration ranges. The average concentration and percentage of the fine mass for the measured constituents in the two concentration ranges are shown in Figure 7b. In the high sulfate concentration regimes the <u>absolute</u> concentration of all measured species increased. However, the <u>relative</u> concentration of both total and elemental carbon decreased. On the other hand the relative amounts of sulfate ion and ammonium ion increased in the high sulfate regime. The presence of secondary organic carbon aerosol is detectable in the high sulfate regime during the summer from the organic carbon to elemental carbon ratio [OC/EC (high sulfate, summer) = 5.2; OC/EC (low sulfate, summer) = 3.8]. However, in the winter sampling there was no detectable secondary organic carbon from this ratio [OC/EC (high and low sulfate, winter) = $\sim$1.9].

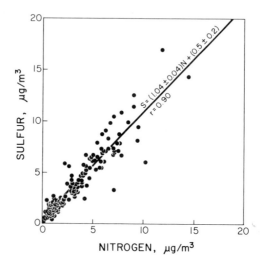

Figure 5.   Plot of S concentration vs. N concentration for St. Louis

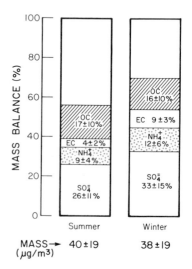

Figure 6.   Mass balance of St. Louis fine aerosol broken down into summer and winter sampling periods

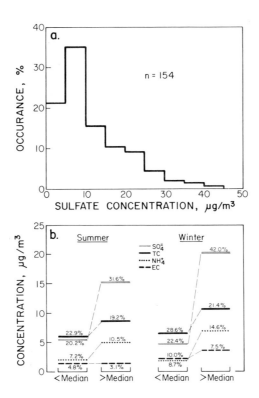

Figure 7. (a) Frequency distribution of St. Louis fine aerosol mass concentration; (b) mass balance of St. Louis fine aerosol divided into high- and low-sulfate concentration regimes for summer and winter. The median sulfate concentration of 8.4 μgm³ was used as the separation point for high- and low-sulfate regimes.

Elemental-Total Carbon Correlations. The relationship
between elemental carbon and total carbon in all samples is
shown in Figure 8a. The data are scattered and not well
correlated (r = 0.63). This is different from the results
reported by Rosen et al. (5), however, our data for 6-h
samples while Rosen examined 24-h samples. In order to more
accurately compare the two data sets we have averaged the
6-h samples to determine the 24-h average concentrations of
elemental and total carbon for the summer and winter, respectively (Figure 8b and c). The winter data show a strong
correlation (r = 0.86) in agreement with the previous work (5);
this correlation is not seen in the summer data (r = 0.14). We
conclude from this that in the winter sampling period in St.
Louis there is very little secondary organic aerosol present.
What is there is covarying with the primary carbon aerosol. On
either time scale in the summer there appears to be significant
amounts of secondary carbon which vary independently of the primary carbon aerosol in St. Louis. This difference in the correlations on different time scales may also imply that diurnal
patterns of primary and secondary carbon aerosol are different.

Automotive Tracers. The work of Cass, Boone and Macias (9)
indicates that in Los Angeles 90% of the elemental carbon
aerosol and 71% of the primary total carbon aerosol is due to
mobile sources. Although a complete carbonaceous aerosol
emission inventory has not been carried out for St. Louis, it is
reasonable to assume that mobile sources of carbonaceous
aerosol are quite important in that airshed as well.

Automotive exhaust is by far the major source of particulate
lead in urban atmospheres (18). In this study we found a strong
correlation between lead and bromine (r = 0.91) during the
winter of 1980, as shown in Figure 9a. This is as expected for
automobile emissions because the principal lead component in
the exhaust is PbBrCl. The Br/Pb ratio of 0.16 is lower than
the stoichiometric value of 0.39 due to the loss of Br from the
aerosol with time. Thus, lead appears to be a good tracer for
auto emissions in the St. Louis aerosol. Interestingly, the
elemental carbon and lead are well correlated (r = 0.84), as
shown in Figure 9b, and yet lead comes mainly from the combustion
of leaded gasoline while the bulk of the elemental carbon comes
from unleaded deisel fuel. The high correlation is most likely
a result of the fairly constant mixture of leaded, unleaded,
and deisel vehicles in the automotive traffic. Total carbon
and lead are also well correlated (r = 0.82) as shown in Figure
9c during this winter period which agrees with the high
correlation between elemental carbon and total carbon (Figure
9d). Thus during the winter period primary emissions from
mobile sources seem to dominate the carbonaceous aerosol. The
least squares fits to the data of total carbon and elemental

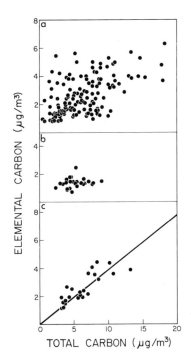

Figure 8. Plot of elemental C concentration vs. total C concentration for particulate samples collected in St. Louis: (a) all samples (6 h); (b) 24-h averages for summer samples; (c) 24-h averages for winter samples.

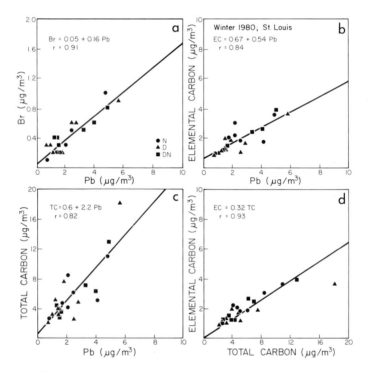

Figure 9. Plots of aerosol constituents in St. Louis, winter 1980: (a) Br vs. Pb; (b) elemental C vs. Pb; (c) total C vs. Pb; and (d) elemental C vs. total C for particulate samples collected during the day (D), during the night (N), and from 3 to 9 pm (DN).

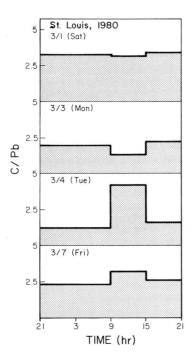

Figure 10.   Plot of the aerosol C/Pb
ratio as a function of time for 4 days in
St. Louis

carbon versus lead have positive intercepts which suggest the
presence of a small but not negligible non-mobile source of
these carbonaceous particles.

The carbon to lead ratio in St. Louis on four winter days
is plotted as a function of time of day in Figure 10. Far less
variation can be seen in comparison to the LA data plotted in
Figure 2 although an increase in the C/Pb ratio in St. Louis
is seen at midday on two days. The C/Pb ratio is also much
lower in the St. Louis data. This supports the previous conclusion that primary carbon is dominant in St. Louis in the
winter sampling period. It also subbests that more secondary
carbon aerosol is present in LA than in St. Louis.

Conclusions

In this paper we have shown that there is no simple answer
to the question posed in the title of this paper. Primary
carbon particles dominate the carbonaceous aerosol under certain
conditions while substantial secondary carbon may be present at
other times. However, the importance of secondary carbon contributions is much less obvious when 24-h samples are examined.
With shorter time averaged samples (e.g. 6-h or less) the
increase in secondary carbon formation can be more easily
detected. Secondary carbon appears to be more important in the
summer rather than winter, in the afternoon rather than the
early morning, and in LA rather than St. Louis. It should be
noted that these conditions of increased secondary carbon
aerosol formation are also more favorable conditions for
photochemical reactions. Our detailed emission inventory
indicates that much more primary carbon exists in the urban
aerosol than was thought previously. This is in agreement
with the data. Our analysis shows that even on the very
smoggy days in the ACHEX study there were times when primary
carbon dominated the carbonaceous aerosol.

Finally, the use of aerosol lead and elemental carbon
have been shown to be useful tracers for primary carbonaceous
aerosol. If these data were combined with information on the
detailed speciation of the organic carbon particles, a rather
complete understanding of the carbonaceous aerosol would be
possible.

Acknowledgements

We are grateful for the help and encouragement of Prof.
Sheldon K. Friedlander of UCLA and Prof. Glen R. Cass of
California Institute of Technology. This work was supported
in part by the USEPA, Inorganic Pollutant Research Branch
under Grant No. R 806005 to Washington University. A portion

of this work was supported by the USEPA under Grant No. R 802160 to Cal Tech while one of us (E.S.M.) was there on sabbatical leave.

Literature Cited

1. Lewis, C.W.; Macias, E.S. Composition of Size-Fractionated Aerosol in Charleston, West Virginia. Atmos. Environ. 1980, 14, 195.

2. Proceedings, Conference on Carbonaceous Particles in the Atmosphere. T. Novakov, Ed., Report LBL-9037, Lawrence Berkeley Laboratory, Berkeley, CA, 1979.

3. Proceedings of the Symposium on Particulate Carbon: Atmospheric Life Cycle. Wolff, G.T.; Klimisch, R.L., Eds., in press.

4. Gartrell, G. Jr.; Heisler, S.L.; Friedlander, S.K. Relating Particulate Properties to Sources: The Results of the California Aerosol Characterization Experiment. In The Character and Origin of Smog Aerosols, Hidy, G., Ed., J. Wiley, New York, 1980, 665.

5. Rosen, H.; Hansen, A.D.A.; Dod, R.L.; Novakov, T. Soot in Urban Atmospheres: Determination by an Optical Attenuation Technique. Science 1980, 208, 741.

6. Hidy, G.M.; Mueller, P.K.; Grosjean, D.; Appel, B.R.; Wesolowski, J.J. The Characterization and Origins of Smog Aerosols, John Wiley, New York, 1980.

7. Appel, B.R.; Colodny, P.; Weslowski, J.J.; Analysis of Carbonaceous Materials in Southern California Atmospheric Aerosols. Environ. Sci. Technol. 1976, 10, 359.

8. Grosjean, D.; Friedlander, S.K. Gas-Particle Distribution Factors for Organic and Other Pollutants in the Los Angeles Atmosphere. J. Air Pollution Control Assoc. 1975, 25, 1038.

9. Cass, G.R.; Boone, P.M.; Macias, E.S. Emissions and Air Quality Relationships for Atmospheric Carbon Particles in Los Angeles. Proceedings of the Symposium on Particulate Carbon: Atmospheric Life Cycle. Wolff, G.T.; Klimisch, R.L. Eds., in press.

10. Hidy, G.; Appel, B.; Charlson, R.J.; Clark, W.E.; Friedlander, S.K.; Giauque, R.; Heisler, S.L.; Mueller, P.K.; Ragaini, R.; Richards, L.W.; Smith, T.B.; Waggoner, W.; Wesolowski, J.J.; Whitby, K.T.; White, W. Characterization of Aerosols in California (ACHEX). Rockwell International,

Final Report SC52425FR to the Air Resources Board of
California, 1975.

11. Macias, E.S.; Husar, R.B. Atmospheric Particulate Mass
Measurement with Beta Attenuation Mass Monitor. Environ.
Sci. Technol. 1976, 10, 904.

12. Macias, E.S.; Radcliffe, C.D.; Lewis, C.W.; Sawicki, C.R.
Proton Induced γ-Ray Analysis of Atmospheric Aerosols for
Carbon, Nitrogen, and Sulfur Composition. Anal. Chem.
50, 1120.

13. Delumyea, R.G.; Chu, L.C.; Macias, E.S. Determination of
Elemental Carbon Component of Soot in Ambient Aerosol Samples. Atmos. Environ. 1980, 14, 647.

14. Cahill, T.A. Elemental Analysis of Environmental Samples.
In New Uses of Ion Accelerators, J. Ziegler, Ed., 1975,
Plenum, NY, p. 1.

15. Stevens, R.K.; Dzubay, T.G.; Russwurm, G.; Rickel, D.
Sampling and Analysis of Atmospheric Sulfates and Related
Species. Atmos. Environ. 1978, 12, 55.

16. Delumyea, R.G.; Macias, E.S.; Cobourn, W.G. Detection of
the Presence of Ambient Acid Sulfate Aerosols from the
Sulfur/Nitrogen Ratio. Atmos. Environ. 1979, 13, 1337.

17. Husar, R.B.; Patterson, D.E. Regional Scale Air Pollution:
Sources and Effects. Annals N.Y. Acad. Sci. 1980, 338, 399.

18. Committee on Biological Effects of Atmospheric Pollutants.
Airborne Lead in Perspective. National Academy of Sciences,
National Research Council, Washington, D.C., 1972.

RECEIVED April 24, 1981.

Comparisons Between Size-Segregated Resuspended Soil Samples and Ambient Aerosols in the Western United States

T. A. CAHILL, L. L. ASHBAUGH, R. A. ELDRED, P. J. FEENEY,
B. H. KUSKO, and R. G. FLOCCHINI

Air Quality Group, Crocker Nuclear Laboratory and Departments of Physics and
Land, Air, and Water Resources, University of California—Davis,
Davis, CA 95616

Soil derived particles generally form the major
fraction of aerosol mass measured in remote, arid
sites in the western United States, based upon results from the 40 station EPA/UCD Western Fine Particle monitoring network. Details of the mechanisms that cause the observed soil mass, size profiles, and chemistry are complicated by the strong
dependence of composition on soil particle size.
Changes of an order of magnitude or more can be
found in the elemental ratios of gross constituents
as a function of size at a single site. Site to
site differences can be equally large. In order to
aid in the interpretation of aerosol data, and especially distinguish between soils and fly ash, o-
ver 100 soil samples were taken at or near aerosol
collection sites in 8 mountain states, chosen so as
to represent likely sources for wind blown soil
near the samplers. Each sample was seived into 5
fractions under dry conditions. The finest fraction was then resuspended in an air stream and injected into a 5 stage Lundgren rotating drum impactor with cut points at 25,8,3, and 0.8 microns
equivalent aerodynamic diameter. Each stage was
analyzed by particle induced x-ray emission (PIXE)
for elements sodium and heavier. Coarse fractions
were poorly correlated with the local aerosol mass,
and bulk analyses were widely different from the
aerosols and the fine particle modes in the resuspended soils. Relatively minor chemical shifts
were seen in the 2.5 to 15 micron size range, but
finer aerosols and soils differ strongly from crustal averages.

0097-6156/81/0167-0269$05.00/0
© 1981 American Chemical Society

In most remote areas of the western United States, the mass
of atmospheric particles in the less than 15μm diameter size range
is dominated by soil-like particles rich in silicon, aluminum,
calcium, iron, potassium, titanium, manganese, with sodium, strontium, and zirconium also present in minor amounts. Even in the
less than 2.5μm fraction, soil-like aerosols form a major, occasionally dominant, component in that size range. Yet, it is precisely the sub-2.5μm size range that is implicated in visibility
degradation, long range transport, and pulmonary deposition, and
thus its importance is great. Detailed studies of such soil-like
particles, designed to identify natural and anthropogenic sources,
are faced with a number of problems. Some fraction of the soil,
though natural in origin, has been resuspended by man's activities, while other soil surfaces have been disturbed and thus contribute disproportionately to wind suspended soil mass. Other
sources, such as coal fly ash, may resemble native soils in gross
composition, making source identification difficult without trace
chemistry, size profiles, and/or morphological information. Finally, individual components of the soil-like elements can be generated by industrial processes, open fires, mining operations,
etc., and these must not be relegated to soil sources. Thus, es-
tablishment of the true natural soil aerosol, identification of
direct and indirect anthropogenic contributions to soil-like aerosols, and connection of both to effects such as visibility degradation are primary goals for understanding air quality in the arid
west.

Experimental Procedures for Ambient Aerosols

The aerosol data used in this paper were collected as part of
the sampling program of the Western Fine Particle Network (2).
The network was established by the Air Quality Group at the University of California, Davis, and the Las Vegas Environmental Monitoring and Support Laboratory of the U.S. Environmental Protection Agency. It covers the states of Montana, North and South Dakota, Wyoming, Utah, Colorado, Arizona, and New Mexico with a relatively uniformly spaced grid of 40 sites (Figure 1). Many of the
sites are in class I areas such as national parks and monuments.
All are remote from local anthropogenic sources. At each site, a
dichotomous stacked filter sampler (2,3) collects samples in two
size ranges; 15μm to 2.5μm, and 2.5μm to 0 μm; based on the fil-
tration properties of Nuclepore filters (4). The units are operated for two 72 hour periods each week. The samples are analyzed
at Davis gravimetrically for mass, and by particle induced x-ray
emission (PIXE) for elemental content (5).
Additional instrumentation provides data on meteorology, visibility, gaseous pollutants, and more detailed particle information at Zion, Yellowstone, Canyonlands, and Carlsbad National

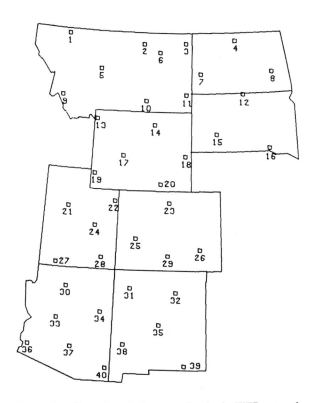

Figure 1. Sites of particulate samplers in the WFP network

Parks, and Cedar Mountain, Utah. The Canyonlands and Zion National Park sites were maintained from a previous study (6) to provide historical data continuity from October, 1977. At these two
sites, and at Yellowstone National Park, a multiday impactor provides daily samples in three size ranges; 15µm to 2.5µm, 2.5µm to
0.5µm, and <0.5µm. The Zion site operates only the multiday impactor. At the other four sites, a dichotomous virtual impactor
(7) operates side-by-side with the stacked filter samplers. Particles are collected in the same size ranges and on the same time
schedule for both samplers. Additional instruments operated for
special studies included ten stage impactors (8) for detailed size
information and single stage streaker samplers (9) for two-hour
time resolution. All particulate samples were analyzed for elemental content by PIXE, and when possible, gravimetrically for
mass (10). The multiday impactor and a prototype stacked filter
sampler were used with the Davis PIXE and x-ray fluorescence (XRF)
analytical systems in the EPA/DOE intercomparison test in Charleston, West Virginia (11).

Ambient Aerosol Results

    In the aerosol data collected by the WFP network from Octo-
ber, 1979 through May, 1980, six predominately coarse elements
consistently correlate with each other: Si,Al,K,Ca,Fe, and Ti.
For the coarse particles at any site, the correlation coefficient
between any pair is typically greater than 0.90. The composition
of this airborne soil is reasonably close to those of standard
soil types. This is shown in Table I, which lists ratios to iron.

TABLE I. COMPOSITION OF AIRBORNE SOIL VS. AVERAGE
CRUSTAL COMPOSITION. VALUES RELATIVE TO IRON

	Mean Airborne (all sites)	Range Airborne (2/3 sites)	Igneous, Sediment, shale*	Sandstone*	Limestone*
Si	6.4	5.1 -7.4	5.8	37	6.4
Al	2.0	1.7 -2.8	1.7	2.6	1.1
Ca	2.0	1.2 -2.3	0.7	4.0	81.
K	.72	.57- .84	.56	1.1	.73
Ti	.16	.11- .25	.10	.15	.10

* From reference (12)

Included is the mean for all sites and the range encompassing the
mean values at 2/3 of the sites. The silicon composition appears
to be less than expected from average rock; at 1/3 of the sites,
the Si/Fe ratio was less than the 5.8 value for igneous, sediment,
and shale.
    The largest range in elemental composition in both the

airborne samples and in the average rock is that exhibited by calcium. At a single site, the Ca/Fe ratio exhibits little variation
with time or particle size. However, at some sites, the airborne
soil is richer in calcium, producing a pronounced spatial variation. For example, the mean Ca/Fe ratio for coarse particles
ranges from 4 to 7 at three calcium rich sites, as opposed to low
means of less than 1.2 at eight sites. Comparing the spread of
average ratios for each site, the standard deviation of the site
averages divided by the mean of the site averages is twice as
large for Ca/Fe (60%) as it is for K/Fe (30%), Al/Fe (30%), or
Si/Fe (22%). The fact that this spatial variation holds even for
fine soils indicates that the soil component of the ambient aerosol originates on a scale smaller than the mean distance between
adjacent sites (140 miles).

In order to consolidate the separate concentrations of soil
elements, a 'soils' concentration was calculated by summing the
elemental concentrations plus their presumed oxide concentration.
These oxide forms are: $Al_2O_3, SiO_2, K_2O, CaO, TiO_2$, and Fe_2O_3. For
the coarse particles, this soils concentration accounted for 60%
of the coarse gravimetric mass, with a correlation coefficient between the two of 0.90. For the fine particles, soils accounted
for only 25% of the fine gravimetric mass, and the two were poorly
correlated.

The aerosol data were examined for spatial and temporal
trends in the soil and gravimetric concentrations. Figure 2
shows the spatial distributions of the average concentration of
the site from October, 1979 through May, 1980. The variations in
the coarse fraction are somewhat greater than those for the fine
fraction. Comparing the ratio of the standard deviation of the
site averages to the mean concentration for all sites, the scatter
for the coarse soil is 60% larger than that of the fine soil (.51
vs. .32). This suggests that the coarse concentrations are more
affected by local sources than are the fine concentrations.

A pronounced seasonal variation is observed in the coarse and
fine fractions of both mass and soils, as illustrated in Figures
3 and 4. Shown are the mean values for each time period for the
northern and southern halves of the network. One significant
trend is the sharp decrease in coarse and fine soil and coarse
mass during the winter months. The duration of this minimum is,
however, different for the fine soil: the fine soil drops sharply in mid-October, while the coarse soil does not drop until January. The coarse soil concentration increases during mid-Decem-
ber, yet there is no corresponding fine soil increase. Also, the
minimum for fine soil is more pronounced in the north than in the
south, perhaps due to increased snow cover at the northern sites.
Finally, these plots show the strong correlation between coarse
soil and coarse mass and the considerably weaker correlation between fine soil and fine mass. It is obvious that most of the
winter episodes of fine mass are not associated with fine soil.

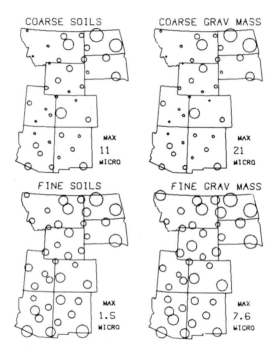

Figure 2. Profiles of fine (< 2.5 μm) and coarse (2.5 μm < D < 15 μm) soil particles and gravimetric mass data from the WFP network from October, 1979 through May, 1980

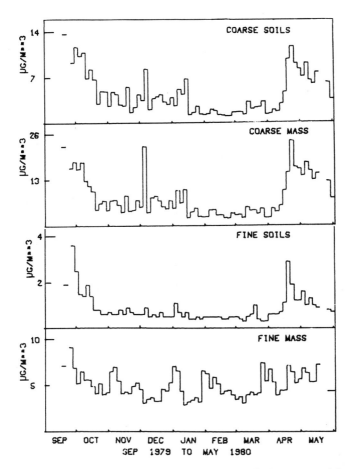

*Figure 3.   Time plots of soils and gravimetric mass for both coarse and fine frac-
tions averaged over the 20 northern sites of the WFP network (3-day mean)*

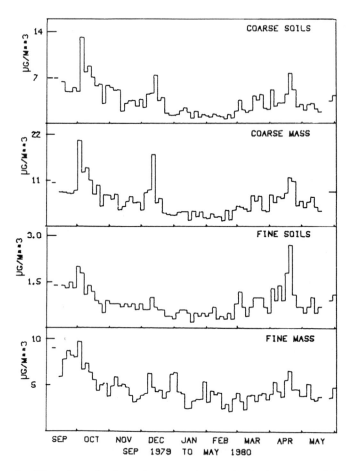

*Figure 4.   Time plots of soils and gravimetric mass for both coarse and fine frac-
tions averaged over the 20 southern sites of the WFP network (3-day mean)*

The temporal variation of airborne soil is shown in even greater detail in the daily concentrations of coarse silicon (3.6 to 15μm) at three multiday sites for 1978 (Figure 5). The concentrations are much lower in winter (mid-September to mid-March) than in summer. Also, the concentrations are dominated by episodes of 4 to 5 days duration. These episodes are somewhat more frequent in summer than in winter (4 per month vs. 2 per month). The fraction of soil aerosol in the fine size range exhibits a distinct spatial variation, as illustrated in a contour plot for iron (Figure 6). The soil aerosols are coarsest in the northern plains (eastern Montana and Wyoming, North and South Dakota), while they are finest in Utah and far western Montana.

Experimental Procedures for Soil Samples

In order to understand the behavior of the soil-like aerosols, samples of exposed surface soils were collected in the vicinity of each of the sites. Additional samples were obtained from locations in the Ute Indian area, for a total of 110 soil samples. The samples were collected by first selecting a location representative of the surrounding area. Factors considered at each site included ground cover, agricultural practices, and a judgement of the ability of the surface to be resuspended. An area approximately 15 cm square was removed to a depth of 2-3 cm and transported to Davis in a plastic zip-lock bag or a nalgene bottle.

The resuspension system was developed to compare fly ash stack samples to bulk samples of hopper material (13). As Figure 7 illustrates, the resuspended hopper material is very similar to the stack samples. The system is used here in an attempt to obtain a soil sample more representative of that suspended in the atmosphere than is the bulk soil sample.

The samples were stored at Davis under low humidity prior to fractionation and resuspension. Approximately 200 grams of sample was sieved into five fractions. The amount in each fraction was weighed to determine the gross particle size distribution. Approximately 1 gram of the fraction which passed through a 50μm mesh was suspended in a low velocity (~30 cm/s) jet, transported 30 cm in a laminar flow at approximately 100 cm/sec. and collected in a five stage Lundgren impactor (Figure 8). The drums of the impactor were covered with mylar and coated with Apiezon L grease. Particles were collected in size ranges of ~50 to 25μm, 25 to 8μm, 8 to 3 μm, 3 to 0.8 μm, and less than 0.8 μm. The drums of the first two stages were rotated to prevent collecting more than a monolayer of material. Since the size distribution reached a minimum on stages 3 and 4, those drums were not rotated. This procedure provided good analytical sensitivity on fine particle stages, and avoided overloading the coarse particle stages.

Results of Soil Analysis

The fraction of soil mass that could be resuspended into

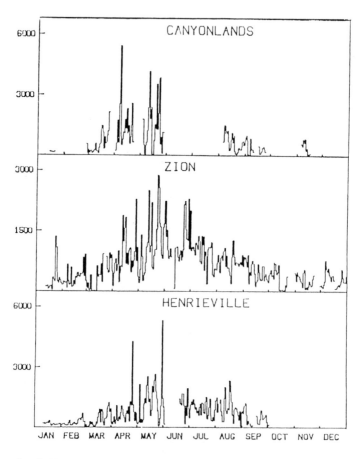

Figure 5.   *Daily concentrations of Si in the coarse size range (15–2.5 μm) at 3 sites in Utah for 1978*

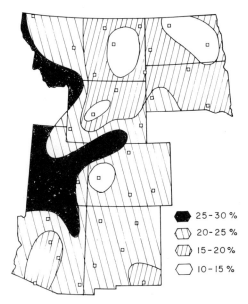

Figure 6. Spatial distribution of the fraction of Fe in the fine range (< 2.5 μm) compared to total Fe (< 15 μm) for aerosol samples collected from October, 1979 to May, 1980

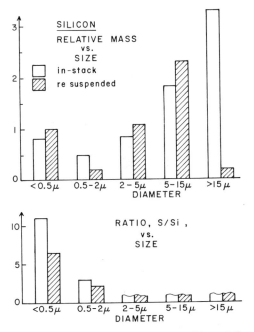

Figure 7. Comparison of mass and elemental composition of fly ash sampled in the stack of a coal-burning power plant, and bulk hopper fly ash, collected while warm on the same day, later resuspended at Davis

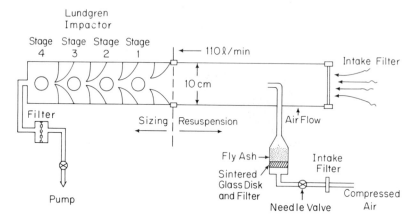

Figure 8.   Schematic of the fly ash resuspension and sizing system used for dusts (University of California—Davis)

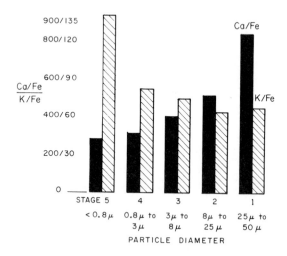

Figure 9.   Ratio of Ca/Fe and K/Fe by size category in resuspended soil from Site 25, Delta County Airport, CO

particles smaller than 15μm was compared to the concentration of ambient soil aerosols measured at each site. If the major factor in aeolian resuspension of soils is associated with the fraction of fine soil particles, then there should be a high correlation between ambient levels and fine soil fractions. However, only a very low correlation between the two was observed. This indicates that other factors such as soil moisture, vegetative cover, local traffic, and wind patterns are too variable and complex to allow prediction of airborne soil mass from the resuspended fraction.

The elemental composition of the individual samples exhibits considerable variation with particle size, especially between the coarsest and finest,   although little variation is observed between the 0.8μm to 3μm size range and the 3μm to 8μm size range. These two ranges correspond closely to the two ranges collected by the stacked filter sampler. Figure 9 shows the Ca/Fe and K/Fe ratios for site 25, which exhibited large variations in both of these ratios versus particle size. Other sites exhibited different variations, so that on the average, very little elemental shift was observed with particle size, for resuspended soils.

Figure 10 is a histogram of the K/Fe and Ca/Fe ratios for each site. The K/Fe ratio exhibits a moderate variation typical of most ratios, while the Ca/Fe ratio exhibits a large variation. Figure 10 also indicates the ratios for the mean of the ambient aerosol measurements and for various rock types. When calculating enrichment factors for aerosol measurements, it is extremely important to choose the reference ratio from a knowledge of the local soil and rock types. Otherwise, errors of an order of magnitude or greater are possible. Figure 11 is a scatter diagram of the Ca/Fe ratio for soils vs. aerosols. For most sites, the ratios are identical, within the ±30% variance,   indicated by the dashed lines. With the exception of K/Fe, most elemental ratios for resuspended soils are similar to the ratios observed in the ambient aerosol. The K/Fe ratio is generally higher in ambient aerosols, suggesting an alternate source of potassium. Other studies (14) have shown that potassium is a good tracer of smoke from forest fires and agricultural burning, indicating that a smoke impact may be present in the WFP aerosol network. This hypothesis is given further support by a comparison of elemental ratios on the fine and coarse stages of the WFP samples. Figure 12 shows the fine/coarse ratio of Si/Fe, T/Fe, Ca/Fe, and K/Fe. Also shown are the ratios of the 0.8μm to 3μm size range to the 3μm to 8μm size range for soil samples. Only the K/Fe ratio for ambient aerosols shows a difference between the coarse and fine stages, being higher in the fine stage for airborne particles.

Summary and Conclusions

1.   The average aerosol mass and soil aerosol in the coarse size fraction exhibit strong spatial variations over the WFP network. The average mass and soil aerosol in the fine fraction exhibit less spatial variation.

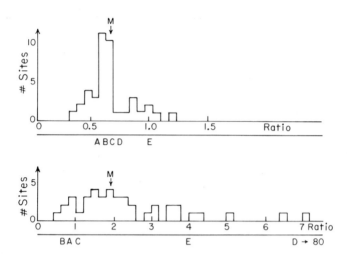

Figure 10.   Ratios of Ca/Fe (bottom) and K/Fe (top) at WFP sites. The letters indicate ratios for WFP aerosol samples (M), igneous rock (A), shale (B), sediment (C), limestone (D), and sandstone (E).

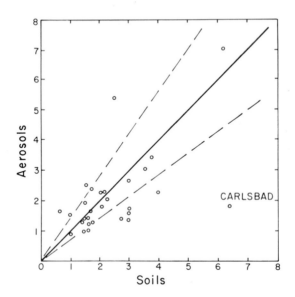

Figure 11.   Ca/Fe scatter plot of resuspended fine soils vs. ambient aerosols for all WFP sites ((– – –) a deviation of ±30% from equal values (———)). The Carlsbad site is located on a limestone outcrop which does not affect the ambient aerosol ratio.

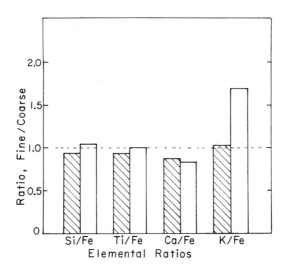

Figure 12. Histogram of average elemental ratios of fine and coarse resuspended soils (▨) and fine and coarse airborne particles (□) at WFP network sites. Excess fine airborne K, not present in soils on the average, is attributed to smoke.

2. Seasonal variations are observed in both the coarse and
    the fine fractions of ambient aerosol mass and soil aer-
    osol. The seasonal trends are dominated by episodes of
    4-5 day duration.
3. The composition of soil aerosol, as indicated by elemen-
    tal ratios, varies spatially. Calcium exhibits the
    greatest spatial variation relative to other soil elements.
    At most sites, little temporal variation exists.
4. The size distribution of soil aerosol varies spatially,
    being relatively coarser at the great plains sites in
    Montana and the Dakotas, and finer in Utah and far
    western Montana.
5. There is a low correlation between the concentration of
    the soil aerosol at a site and the fraction of soil less
    than 15μm.
6. The relative amounts of calcium in the resuspended soils
    has a strong spatial variation. The relative amounts of
    other soil elements exhibit a less pronounced spatial
    variation.
7. At each site, the relative composition of the resuspended
    soil varies strongly with particle size. The variation
    is different for each site, however, so that the results
    from one site cannot be generalized to another.
8. Except for potassium, the composition of the soil aerosol
    at individual sites resembles the composition of the
    resuspended soil at the same site. The excess potassium
    is attributed to smoke sources.

Literature Cited

1. Flocchini, R.G.; Cahill, T.A.; Pitchford, M.L.; Eldred, R.A.;
    Feeney, P.J.; Ashbaugh, L.L. "Composition of Particles in the
    Arid West"; Atmos. Environ., 1980, in press.

2. Cahill, T.A.; Ashbaugh, L.L.; Barone, J.B.; Eldred, R.A.;
    Feeney, P.J.; Flocchini, R.G.; Goodart, C.; Shadoan, D.J.;
    Wolfe, G.W. "Analysis of Respirable Fractions in Atmospheric
    Particulates Via Sequential Filtration"; APCA Journal 1977,
    27(7), 675.

3. Cahill, T.A.; Eldred, R.A.; Barone, J.B.; Ashbaugh, L.L. "Am-
    bient Aerosol Sampling with Stacked Filter Units"; Federal
    Highway Administration Report No. FHWA-RD-78-178, 1979.

4. Spurny, K.R.; Lodge, J.P., Jr.; Frank, E.R.; Sheesley, D.C.
    "Aerosol Filtration by Means of Nuclepore Filters: Structural
    and Filtration Properties"; Environ. Sci. Technol., 1979, 3,
    453.

5. Cahill, T.A. "Ion Excited X-ray Analysis of Environmental Samples"; in New Uses of Ion Accelerators, Ziegler, J., ed.; Plenum Press; New York, 1975, 1-72.

6. Flocchini, R.G.; Cahill, T.A.; Ashbaugh, L.L.; Eldred, R.A.; Pitchford, M.L. "Seasonal Behavior of Particulate Matter at Three Rural Utah Sites"; Atmos. Environ., 1980, in press.

7. Dzubay, T.G.; Stevens, R.K. "Ambient Air Analysis with Dichotomous Sampler and X-ray Fluorescence Spectrometer"; Environ. Sci. Technol., 1975, 9, 663.

8. Johansson, T.B.; Van Grieken, R.E.; Nelson, J.W.; Winchester, J.W. "Elemental Trace Analysis of Small Samples by Proton Induced X-ray Emission"; Anal Chem, 1975, 47, 855.

9. Nelson, J.W. "Proton Induced Aerosol Analyses: Methods and Samplers"; in X-Ray Fluorescence Analysis of Environmental Samples, Thomas Dzubay, ed., Ann Arbor Science Publishers, Ann Arbor, 1977, 19-34.

10. Engelbrecht, D.; Cahill, T.A.; Feeney, P.J. "Electrostatic Effects on Gravimetric Analysis of Membrane Filters"; APCA Journal, 1980, 30(4), 391.

11. Camp, D.C.; Van Lehn, A.L.; Loo, B.W. "Intercomparison of Samples Used in the Determination of Aerosol Composition"; U.S. Environmental Protection Agency Report EPA-600/7-78-118, 1978.

12. Weast, R.C.; Selby, S.M.; Eds. "Handbook of Chemistry and Physics, 48th Ed.", Chemical Rubber Company, Cleveland, 1967, p. F135.

13. Cahill, T.A.; Ashbaugh, L.L. "Size/Composition Profiles of Resuspended Fly Ash"; in Environmental and Climatic Impact of Coal Utilization, Jag J. Singh and Adarsh Deepak, Eds.; Academic Press, New York, 1980, 569-573.

14. Lyons, C.E.; Tombach, I.; Eldred, R.A.; Terraglio, F.P.; Core, J.E. "Relating Particulate Matter Sources and Impacts in the Willamette Valley During Field and Slash Burning"; presented at 72nd Annual Air Pollution Control Association Meeting, June 24-25, 1979.

RECEIVED March 10, 1981.

Aerosol Composition in Relation to Air Mass Movements in North China

JOHN W. WINCHESTER, MICHAEL DARZI, and ALISTAIR C. D. LESLIE
Department of Oceanography, Florida State University, Tallahassee, FL 32306

WANG MINGXING, REN LIXIN, and LÜ WEIXIU
Institute of Atmospheric Physics, Chinese Academy of Sciences, Beijing, People's Republic of China

The elemental composition of aerosol particles has been investigated, as a function of both particle size and time, at a nonurban mountain location 110 km NE of the city of Beijing in order to determine relationships of composition with large scale air mass movements. The study period, 16-21 March 1980, was selected so as to follow the end (15 March) of much of the residential space heating by coal combustion in Beijing, thus reducing air pollution, and to precede a season of dust storms and the monsoon shift to southerly air flow from urbanized eastern China. Earth crustal aerosol and marine aerosol components could be resolved from some pollution aerosol by use of both meteorological and chemical data. Concentrations of the crustal component, mainly in a coarse particle mode (Al, Si, K, Ca, Ti, Mn, Fe) but with some fine mode admixture (Zn, Mn), varied in time. The concentration maxima and minima were approximately synchronous with those of the partial pressure of water vapor, P_{H_2O}, signifying aerosol concentration changes with air masses as they crossed the sampling site. A period of high concentrations of coarse particle Cl corresponded to an incursion of marine air. Fine mode S, K, Cl, and Pb, also observed, are attributed mainly to a pollution component formed from coal combustion and other processes in the Beijing area.

The Beijing area in north China is well situated for investigating relationships between the elemental composition of atmospheric particulate matter and air mass movements. The municipality of Beijing (Beijing shi) has a population of 7 million, half of whom live in the city proper, and pollution sources of trace gases

0097-6156/81/0167-0287$05.00/0
© 1981 American Chemical Society

and aerosols are numerous. During winter coal combustion for
space heating is a major pollution source type, but this offici-
ally ends on 15 March in Beijing. Some steel and chemical industries, located southwest and southeast of the city, produce additional air pollution. North of the city the population is mainly
engaged in agriculture, and north of the Great Wall (e.g. at Bada-
ling 50 km north of the center of Beijing) the population density
is quite low owing to the steppe and desert lands. The ocean lies
some 100 km east of the city, permitting incursions of marine air.
During the cold months, air flow is mainly from the north and west,
bringing relatively clean continental air from Mongolia and Sibe-
ria over the city. During the warm months, air flow is mainly
from the south and east, bringing warm and humid air and heavy
monsoon rainfall to the city. On both a seasonal and a shorter
term basis Beijing experiences contrasting weather and air flow
characteristics, reflected in contrasting atmospheric chemical
properties and aerosol particle composition. The present investigation was conducted to document these in a preliminary way in
order to lay a groundwork for future more detailed studies of regional aerosol chemistry and long range air pollution transport.

Meteorological Conditions During Aerosol Sampling

      For this investigation a program of aerosol sampling and me-
teorological measurements was planned at a nonurban site 110 km NE
of the center of Beijing, commencing on 9 March and continuing
until 3 April 1980. This period included the last week of the
official winter space heating season in Beijing (to 15 March) and
the last 3 weeks of the official heating season in rural areas
north of the Great Wall (to 31 March). The present paper is based
primarily on measurements made 16 to 21 March. The site, an astronomical observing station of the Chinese Academy of Sciences near
Xinglong, is located in a mountainous agricultural area of the Yan
Shan range, latitude 40°23'N and longitude 117°30'E, at 960 meters
elevation just east of the border between Beijing shi and Hebei
province and about 30 km north and east of parts of the Great Wall.
Wuling Shan, 2100 meters elevation and about 25 km north of the
site, was the highest peak in the vicinity and could be clearly
seen on some days but was obscured by regional haze on other days
during the measurement program described here. Except for a small
coal burning heating plant and a kitchen near the site, local air
pollution sources were virtually absent, and the aerosol characteristics reported here are considered to be essentially regional
with little local influence.
      Measurements made during the second week, 16-21 March 1980,
are especially relevant to the chemical characterization of air
masses. During this period meteorological conditions can be summarized as follows, based on weather information from the Institute
of Atmospheric Physics and twice daily surface and upper level
weather maps published by the Japan Meteorological Agency.

Atmospheric conditions over the sampling site were variable, and several different air masses passed by. On 16 March the weather map of the northern hemisphere shows two high pressures, one over Lake Baikal and the other over South Korea, so that the sampling site was just in between; the air was calm, cloudy, and conducive for pollutants to build up. Later the north high pressure moved southwards. By midday of 17 March the sampling site was dominated by high pressure. At that time air flow over the site was mainly from the northwest, the wind speed was 3 meters/second, and the water vapor content was very low. The sky was clear, and vertical dilution was effective.

After the high pressure passed over, a cyclone was formed to the northwest of the sampling site, which caused southeasterly wind and brought warm and moist marine air to the site from late 17 to 18 March. Late on 18 March the marine air mass moved backwards because of strong north high pressure, and a cold front formed and passed over the sampling site shortly after its formation early in the morning of 19 March. After the passing of the cold front, there was a 2-day period of clean continental air mass moving over the sampling site. On 21 March a warm air mass moved northwards again and caused snow by midnight.

Experimental Procedures

The methods employed in this investigation were principally based on aerosol sampling and elemental analysis by particle induced X-ray emission, PIXE, as reviewed by Johansson and Johansson (1). These methods are discussed with specific reference to the present study by Wang et al. (2). Because of the inherent sensitivity of the PIXE method and its suitability for analysis of samples having a few mm^2 area, PIXE-compatible aerosol samplers were employed. These were: (1) time sequence filter samplers, "streakers", which collect an aerosol deposit as a streak along the surface of a 0.4 μm pore size Nuclepore filter strip by means of a continuously moving sucking orifice drawing air at 0.7 liter/minute through the filter while sliding along its smooth back side (3); and (2) single orifice cascade impactors with 8 stages and particle size cuts (for 1 liter/minute flow rate) at 0.25, 0.5, 1, 2, 4, 8, and 16 μm aerodynamic diameter (μmad). (Actual size cuts are slightly greater at the 0.8 liter/minute flow rate used here.) The streakers provide up to 14 days of sample streak (168 mm long) on a single filter which could be analyzed in 4-hour time steps using a 2-mm collimated proton beam. The impactors were fabricated from plastic (PIXE International, Inc., Tallahassee, FL, U.S.A.) and operated for 10 or 12 hour time periods to collect 8 particle size fractions on Vaseline or paraffin coated Mylar film or, for <0.25 μm particles, 0.4 μm Nuclepore filters. The impactors were operated inside special wind speed decelerators, consisting of two concentric perforated cans, in order to increase particle collection efficiencies up to about 15 μmad. (The decelerators were

provided by T.A. Cahill, University of California, Davis, CA,
U.S.A.) Both types of samplers were operated facing downwards
using Brailsford, Inc. (Rye, NY, U.S.A.) vacuum pumps and lead
acid automobile storage batteries. A discussion of sampling and
analysis procedures, impactor reproducibility, and sampler inter-
comparability, together with representative data for the 9-16 March
1980 period is given by Winchester et al. (4). Wang et al. (5)
also present a detailed description of the sampling site and locations of the aerosol samplers at the Xinglong astronomical observing station. PIXE analysis was carried out at Florida State University.

During the aerosol sampling from 9 to 21 March, temperature
(-8° to +10°C) and relative humidity (about 10% to 98%) were recorded continuously on chart paper by thermocouple and hair hygrom-
eter sensing elements calibrated daily by wet and dry bulb mercury
thermometers. Water vapor partial pressure, P_{H_2O}, in millibars
was calculated for hourly intervals from this record. Wind speed
and direction were recorded continuously at the sampling site, and
periodic measurements of barometric pressure (950 to 980 mb) as
well as frequent observations of local weather conditions were made
throughout the sampling program. Additional aerosol particle
counts were made by light scattering instruments principally during
the daytime, and atmospheric transparency was determined by a solar
radiometer during sunny days; the results of these measurements
will be reported elsewhere.

Results

During the 16-21 March 1980 study period the water vapor
pressure, P_{H_2O}, exhibited variability with time at the Xinglong
site which corresponded to the succession of air masses which
passed over the site. Figure 1 shows this variability together
with the patterns of concentration variability of Fe, S, and Cl
collected by the Nuclepore filter of the streaker and determined
by PIXE. Also indicated are approximate times of occurrence of
the different air masses, inferred from both meteorological conditions and chemical characteristics: a period of rather polluted
air (P) on 16-17 March, one with marine air mixed with continental
(M+C) on 17-18 March, then continental air with some pollutants
(C+P) on 18-19 March, followed by rather unpolluted continental
air (C) on 20 March, and finally continental air with rising pollution levels (C+P) on 21 March. Along the top of Figure 1 are given
the midpoints of alternating 10 and 12 hour periods of samples
collected by cascade impactors, discussed further below.

Fe is representative of a group of elements in particulate
matter derived largely from earth crust minerals. Its trace in
Figure 1 shows times of maximum and minimum concentrations which
usually occur within a few hours of a similar pattern in P_{H_2O}. If
we interpret the major variations of P_{H_2O} to be the result of air
mass changes, such as by frontal passages, then large changes in

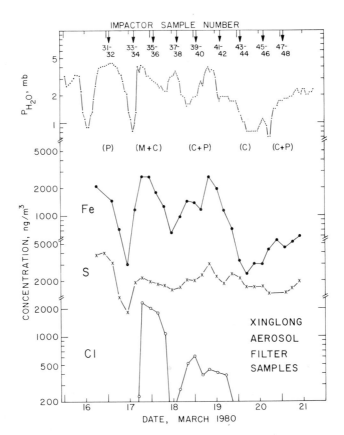

Figure 1. Water vapor pressure (P_{H_2O} in mbars) and aerosol Fe, S, and Cl concentrations (ng/m³) measured using streaker sampler at Xinglong.

Approximate times of polluted (P), continental (C), and marine (M) air masses are indicated based on synoptic weather maps and consistent with aerosol composition measurements. Times of impactor samples, taken in duplicate concurrently with the streaker, are indicated at the top as midpoints of alternating 10- and 12-h sampling periods.

the crustal aerosol concentrations may also represent changes in relatively large scale air mass composition characteristics. We regard the calibration of streaker timing to be accurate to 0.5-1 time step (2-4 hours). Since some of the time differences between corresponding features of the Fe and P_{H_2O} traces are greater than this uncertainty, there may be small but real differences in arrival times of these constituents of successive air masses at the sampling site.

S occurs generally at around 2000 ng/m^3 (higher on 16 March) and with a pattern of variability unlike that of Fe. We believe that aerosol S here is mainly an air pollutant and that its concentration variations are governed by several factors. These include gaseous SO_2 concentrations, the rate of gas-to-particle conversion during transport, and atmospheric mixing processes. The higher initial concentrations may be attributed in part to higher SO_2 pollution source strengths on 15 March, the last day permitted for much of the residential space heating by coal combustion in Beijing, but subsequent air pollution by SO_2 from other utilization of coal in this part of north China still maintains aerosol S levels far above natural continental levels. Thus, aerosol S cannot be considered as characteristic of air mass composition in the same sense as can aerosol Fe.

Cl was generally found on the Nuclepore filter of the streaker sampler at levels very much lower than in cascade impactors operated concurrently. This result, discussed further below, is attributed to volatility of Cl on the filter, such as by chemical reaction with acidic or other reactive agents drawn through the filter. The cascade impactors should have much less tendency for such volatilization, owing to the air passing by, rather than through, the impaction surfaces which collect particles down to about 0.25 μmad. Therefore, the pattern of Cl variability is usually not to be interpreted as representing only changes in aerosol Cl concentration in the atmosphere but instead as the resultant of both the concentration in air and chemical reactivity on the filter. This may be especially true when sampling polluted atmospheres which contain sulfuric acid and other reactive substances. However, in one part of the record of Figure 1, 17-18 March, Cl concentrations on the streaker filter were unusually high and not lower than Cl concentrations made by simultaneous impactor measurements. This period corresponds to an incursion of marine air from the east which was also inferred from meteorological data.

Concentrations of the elements Si, Al, K, Ca, and Ti varied in time with patterns nearly identical to that of Fe. Figure 2 shows that their ratios to Fe are quite invariant and close to the average values for earth crust materials given by Mason (6). The ratio Ca/Fe fluctuates slightly, and Ti/Fe averages somewhat higher on 20-21 March than previously, but principally a terrestrial dust source for all six elements is considered likely in these samples. For Mn and Zn, however, straightforward dispersion of terrestrial dust is not the only significant source process.

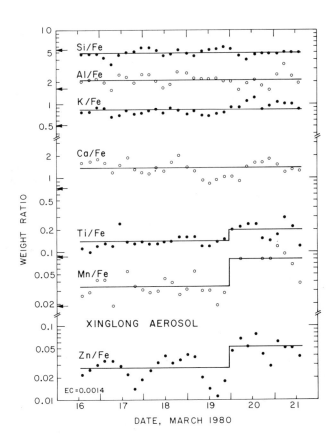

Figure 2.   Weight ratios of elements whose concentrations varied similarly to Fe (Figure 1).

Average ratios for Si, Al, K, Ca, and Ti to Fe approximate earth crust ratios, indicated by arrows at left. These elements occur mainly in coarse particles (Figures 3 and 4). The Mn/Fe ratio is somewhat greater and Zn/Fe is 20 times greater than earth crust ratios. These two elements have appreciable fine-mode concentrations (Figure 4). All 8 elements may occur characteristically in continental air masses.

Figure 2 indicates Mn/Fe to be somewhat above the crustal ratio
through 19 March, and thereafter a marked increase is seen. The
aerosol ratio Zn/Fe averages about 20 times greater than in the
earth crust (somewhat greater on 20-21 March), showing "anomalous"
atmospheric enrichment of Zn first recognized by Rahn (7). Since
particle size distribution measurements, discussed below, show
substantial fine particle concentrations of both Zn and Mn, the
processes for their transfer to the atmosphere must be different
from those for the other six elements of Figure 2. However, their
concentration variations in time still resemble those of Fe shown
in Figure 1 and therefore these elements may also be relatively
large scale characteristics of air masses, in contrast to S where
regional pollution sources and aerosol formation processes must be
important.

The average particle size distributions for four predominantly
crustal elements, Al, Si, Ca, and Ti, are shown in Figure 3. They
are essentially identical. It should be pointed out that the down-
turn of the relative concentrations above 8 μmad (impactor stage 6)
is the combined result of the actual distribution of particle sizes
in the atmosphere and the efficiency with which these very coarse
particles can enter (upward) into the cascade impactor. This efficiency must decrease with increasing particle size and generally
depend on inlet design and wind speed. Nevertheless, it is important to note here that the patterns of the four elements are similar, implying a common aerosol source.

Figure 4 presents particle size distributions for six elements
which differ among themselves and also from those in Figure 3.
Somewhat subjectively, we may identify three patterns in these distributions: (a) A coarse mode, typified by Ca and the other elements of Figure 3, which may represent a terrestrial dust origin.
This mode can account for coarse particle concentrations observed
for Fe, K, Mn, and S. (b) A fine mode with somewhat greater concentrations in the 0.5-1 μmad fraction than in 1-2 μmad particles.
The amounts in this <2 μmad range, in excess of those which can be
attributed to a coarse crustal aerosol tail with the Ca distribution, show similarities in particle size distributions for Zn, Mn,
and possibly Fe. Since the trends shown in Figure 2 point to
these elements being characteristic of large scale air masses,
their fine modes may be principally due to natural processes.
(c) A fine mode with very much greater concentrations in the 0.5-1
μmad fraction than in 1-2 μmad particles. Similar patterns of
this mode are observed for S, K, Pb, and also Cl (cf. Figure 5).
The principal source may be air pollution emissions from coal combustion, suggested by the fact that aerosol S levels are generally
high and vary independently of Fe (Figure 1), Pb was found only in
the most polluted air (16-17 March), and fine particle K and Cl
show some temporal similarities presented in Figure 5. When pollution contributions to the total aerosol are very large, Zn and
other elements with fine modes may also be enhanced by pollution
additions to their natural levels.

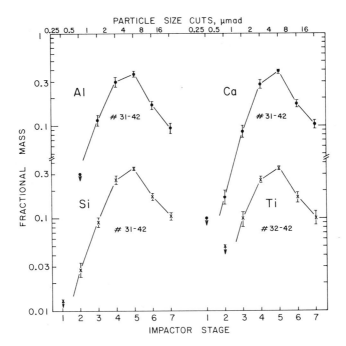

Figure 3. *Particle size distributions of crustal elements Al, Si, Ca, and Ti averaged during March 16–19 (for impactor sample numbers indicated, cf. Figure 1) when concentrations were high enough to provide accurate data in the fine fractions.*

Approximate size cuts between impactor stages are given across the top. No significant concentrations were found on Stage 0, < 0.25 μmad (Xinglong aerosol; means and standard errors of means).

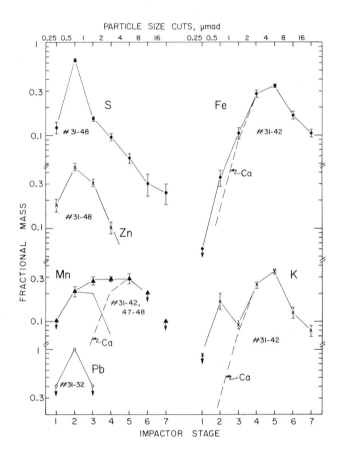

*Figure 4.   Particle size distributions of 6 elements that exhibit fine-mode concen-
trations, as distinct from a coarse-mode typified by Ca.*

The S, Zn, and Pb are found only in a fine-mode centered at 0.5–1 μmad. The Mn and
K can be resolved into both fine- and coarse-modes by reference to Ca, and Fe shows
evidence of a possible fine mode superimposed on a much larger coarse mode similar to
Ca (Xinglong aerosol, March 16–21, 1980; means and standard errors of means).

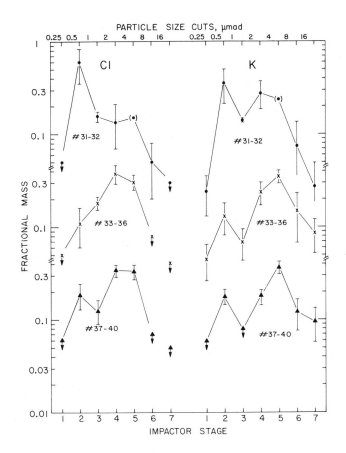

Figure 5. Particle size distributions of Cl and K during 3 time periods (cf. Figure 1) characterized by polluted (P), marine (M), and continental (C) air masses at Xinglong site.

Fine-mode Cl and K may be mainly pollution-derived, coarse-mode Cl typical of marine air, and coarse-mode K typical of continental air (Xinglong aerosol, March 16–19, 1980; means and standard deviations).

Figure 5 provides further evidence that air pollution may contribute significantly to the concentrations of both Cl and K in the fine mode. During the 16-17 March polluted period, which we expect contained some effects of the last days of the Beijing residential space heating season, both Cl and K occurred prominently in fine particles of 0.5-1 μmad. With a change in air mass to one from the east, which should contain marine aerosol, the Cl distribution is mainly coarse as expected from particles formed from seawater droplets. The relative abundance of the fine mode of K is low, although still present. Finally, with a diminishing of the marine air and a shift to more continental air flow, with some pollution aerosol buildup, a fine mode of Cl reappears, and the fine mode of K becomes more prominent. For these two elements, more than is the case for the other elements measured, the interplay of air mass movements and mixing of continental, marine, and polluted air is clearly reflected in changes in aerosol particle size distributions.

Discussion

The week of observations at Xinglong reported here, 16-21 March 1980, was one in which the effect of air pollution on aerosol composition was lessened by the end on 15 March, by official policy, of much residential coal burning for space heating in Beijing. A brief report of results at Xinglong for the week of 9-16 March is published elsewhere (4). After 15 March the larger scale aerosol composition characteristics and their relation to air mass movements can be more readily discerned, as has been described above.

It is of interest to compare the aerosol composition of polluted air on 16-17 March with results of the preceding week. This comparison is given in Table I. Element ratios are presented relative to excess fine particle potassium, K_x, i.e. the fine mode which has been resolved from total K as shown in Figure 4. The fine modes Fe_x and Mn_x were obtained in a similar way.

Two periods could be identified which probably involved considerable air pollution transport to the sampling site, 11-13 March and 15-17 March. The first was sampled as a series of 5 half-day periods and the second as 2 half-day samplings with a change in sampling location on 16 March, from the center of the Xinglong observatory to the western edge (5). In order to test for a possible influence by smokestacks within the observatory, this change was made to a generally upwind location. The three results are shown separately in Table I.

The last two columns show agreement, indicating no apparent dependence of relative elemental composition on sampling location within the observing station or on its own small emissions of smoke. Both columns are in approximate agreement with the 11-13 March period averages, but small differences may be significant: the earlier period is substantially higher in aerosol S relative

Table I.
Xinglong Polluted Aerosol Composition,
Comparison with Preceding Periods

Elements[b]	Weight Ratios, Fine Particles[a]		
	11-13 March[c]	15-16 March[d]	16-17 March[e]
S/K_x	15.8 ± 3.4	6.7	4.8
Cl/K_x	1.2 ± 0.5	0.9	1.0
Fe_x/K_x	0.5 ± 0.2	0.27	0.24
Mn_x/K_x	0.11 ± 0.08[f]	0.04	0.05
Zn/K_x	0.19 ± 0.07	0.13	0.13
Pb/K_x	0.14 ± 0.05	0.08	0.06
Pb/Zn	0.77 ± 0.17	0.7	0.7

[a] The 0.5-2 μmad range from sums of impactor stages 2 and 3, except for Pb/K_x and Pb/Zn on 16-17 March where 0.5-1 μmad, stage 2 data only were used.

[b] Concentrations K_x, Fe_x, Mn_x represent excess over that attributed to a coarse mode with Ca particle size distribution.

[c] Means and standard deviations of 5 sampling periods (7 impactor samples) (4).

[d] Means of duplicate impactor samples (4).

[e] Means of simultaneous samples 31 and 32, start time 1935 on 16 March, stop 0725 on 17 March 1980.

[f] Progressive decrease from 0.22 to 0.025, 11 to 13 March.

to K_x, suggesting a greater degree of SO_2 conversion to H_2SO_4. Mn_x/K_x decreased progressively from highest values on 11 March, suggesting different source types and locations for these elements. Fine mode Fe, Zn, and Pb average slightly greater in 11-13 March samples, relative to K_x, than in 15-17 March samples. The ratio Pb/Zn is quite constant for both periods.

It should be noted that Pb concentrations in this polluted air are about 100 times lower than the aerosol S levels, or are only a few tens of ng Pb/m^3. In the Beijing area the contribution of automotive emissions to atmospheric Pb at Xinglong is expected to be minor owing to the small amounts of leaded gasoline used; privately owned automobiles, trucks, and other gasoline powered vehicles are nonexistent in China. Instead, coal combustion may be the principal source of fine particle Pb as well as K and S. Coal also may contribute to the fine modes of Zn, Mn, Fe, and other metals in the polluted atmosphere. However, the relative elemental compositions of fine mode aerosol particles in polluted and clean atmospheric samples from the Beijing region are quite similar, so that the pollution and natural components are not readily distinguished by composition alone (4).

The natural air mass characteristics of aerosol composition on which the pollution components are superimposed, have been identified in this paper by comparing elemental concentrations and particle size distributions, water vapor pressures, and synoptic weather data. Much of the evidence suggests the following: continental air masses contain elements from terrestrial dust in a coarse particle mode, including Fe, Al, Si, K, Ca, Ti, and Mn, together with additional elements in a fine particle mode generated by natural processes, including Zn and Mn. Marine air adds coarse particle Cl to a continental air mass background. Additional fine particle S, K, Cl, Pb and perhaps Zn and other elements are attributed to pollution, mainly from coal burning. By a judicious choice of sampling time and location and careful attention to meteorological information, it is feasible to resolve the aerosol composition typical of continental and marine air from the pollution aerosol constituents which may be present. The variations of the concentrations of Fe and the other natural constituents are approximately correlated with P_{H_2O} and therefore may serve as chemical indicators of air mass movements in unpolluted areas.

Acknowledgments

The authors are indebted to Bi Mutian, Zhu Wenqin, Wu Zhiyu, Wang Jiajun, and Shen Jianqing and to the staff of the astronomical observing station of the Chinese Academy of Sciences at Xinglong for their assistance with the aerosol sampling program, and to Wang Qingrong for assembling meteorological information used in data interpretation. This work was supported in part by the Institute of Atmospheric Physics, Chinese Academy of Sciences, Beijing, for field measurements and aerosol sampling, by The Florida State

University and the U.S. Environmental Protection Agency for PIXE analysis of aerosol samples, and by the U.S. National Academy of Sciences Committee on Scholarly Communication with the People's Republic of China for a grant to one of us (J.W.W.) as a visiting research scholar in China. The advice and encouragement given to the project by Ye Duzheng and Zhou Xiuji are greatly appreciated.

Literature Cited

1. Johansson, S.A.E.; Johansson, T.B. Analytical applications of particle induced X-ray emission; Nucl. Instr. Meth. 1976, 137, 473-515.
2. Wang Mingxing; Winchester, J.W.; Lü Weixiu; Ren Lixin. Sampling and chemical analysis techniques of atmospheric aerosols (in Chinese); Journal of Environmental Science (Huanjing Kexue) 1980, to be published.
3. Nelson, J.W. Proton-induced aerosol analyses: methods and samplers; in Dzubay, T.G., Ed. "X-ray Fluorescence Analysis of Environmental Samples"; Ann Arbor Science: Ann Arbor, MI, 1977; pp. 19-34.
4. Winchester, J.W.; Wang Mingxing; Ren Lixin; Lü Weixiu; Hansson, H.C.; Lannefors, H.; Leslie, A.C.D. Nonurban aerosol composition near Beijing, China; Nucl. Instr. Meth. 1980, in press.
5. Wang Mingxing; Winchester, J.W.; Lü Weixiu; Ren Lixin. Aerosol composition in a nonurban area near the Great Wall (in Chinese); Scientia Atmospherica Sinica (Daqi Kexue) 1980, to be published.
6. Mason, B. "Principles of Geochemistry"; John Wiley & Sons: New York, 1966; p. 45.
7. Rahn, K.A. "Sources of trace elements in aerosols--an approach to clean air"; Ph.D. thesis, University of Michigan, Ann Arbor, 1977.

RECEIVED March 10, 1981.

Sources of Airborne Calcium in Rural Central Illinois

DONALD F. GATZ, GARY J. STENSLAND, and MICHAEL V. MILLER

Atmospheric Chemistry Section, Illinois State Water Survey, P.O. Box 5050, Station A, Champaign, IL 61820

ALISTAIR C. D. LESLIE

Department of Oceanography, Florida State University, Tallahassee, FL 32305

In a search for sources of alkaline materials in rural air and rain, we have sampled and performed multi-element analyses on ambient particulate matter and potential source materials. Ambient aerosols were sampled daily using single Nuclepore[TM] filters or Florida State University "streakers." Samples of soil and unpaved road materials were also collected and analyzed. The samples were analyzed by various multi-element methods, including ion- and proton-induced X-ray emission and X-ray fluorescence, as well as by atomic absorption spectrophotometry. Visual observations, as well as airborne elemental concentration distributions with wind direction and elemental abundances in aerosols and source materials, suggested that soil and road dust both contribute to airborne Ca. Factor analysis was able to identify only a "crustal" source, but a simple mass balance suggested that roads are the major source of Ca in rural central Illinois in summer.

The sources of particles in the atmosphere are of much current interest. This is true from the viewpoint of both the geo-atmospheric scientist and the pollution control specialist. For the atmospheric chemist or geochemist, the identities and the relative contributions of sources are important for understanding 1) atmospheric element cycles and budgets, and 2) the chemical content of precipitation. The pollution control decision-maker must know the relative contributions of natural and specific man-made sources of toxic materials in order to establish criteria for control regulations.

The objective of this work was to identify calcium sources and their relative contributions in rural central Illinois as a step toward understanding the role of airborne alkaline

0097-6156/81/0167-0303$05.75/0
© 1981 American Chemical Society

materials in local precipitation chemistry. The approach was
one of field collection of ambient aerosols and source materials followed by measurement of their chemical composition.
The data were interpreted using a variety of methods including
examination of element concentration variations with wind
direction, factor analyses, and chemical element balances.

Procedures

Samples of ambient aerosol and source materials were
collected during the summer and fall of 1978 at the Water
Survey's atmospheric chemistry sampling site, 10 km southwest of Champaign-Urbana, Illinois. A map of the immediate
vicinity of the sampling site is shown in Figure 1. The
ground site is a grassy field, measuring 120 x 240 m and
surrounded by corn and soybean fields. The tower used for
sampling is 35 m high and located 360 m to the southeast
of the ground site. Although the two sites are not coin-
cident, we expect conditions on the tower to be representative
of those 35 m above the ground site, except on rare and brief
occasions when the top of the tower is in the plume from a
tractor engaged in tilling or harvesting operations.

Ambient Aerosols. Aerosols were sampled both continuously,
using Florida State University "streaker" samplers (1), and
on a daily basis. The streaker samplers draw air through
a strip of 0.4 μm pore diameter Nuclepore[TM] filter exposed face
downward. The sampler uses a vacuum orifice which moves the
length of the strip of filter at a rate of 1 mm hr^{-1}, thus
providing one week's sample in a "streak" 168 mm long. We
operated one streaker sampler about 1 m above a grassy field
at our ground site (Figure 1) and another on top of the tower
described above. Samples were analyzed in 2-hr time increments
using proton induced X-ray emission (PIXE) (2) at Florida
State University.

Daily 12-hr filter samples were collected on preweighed
37 mm diameter Nuclepore[TM] filters with 0.8 μm pore diameters.
These filters were re-weighed after sampling to obtain total
particle mass. Total mass was used as the basis of the elemental
mass fractions described later, and is known to within a few
percent. One set of filters was collected face down under an
inverted funnel rain shield about 1.5 m above grass. Another
set was collected in a specially-made device we call a vane
sampler. It consists of a plastic (polyvinyl chloride)
tube with a gradual 90° bend and a constricted opening
designed to provide quasi-isokinetic sampling of particles at
wind speeds of about 4 m sec^{-1}. It is provided with a tail
fin and is free to rotate about a vertical axis, so it always
faces into the wind. Elemental analysis of these filters was
performed at Crocker Nuclear Laboratory, University of

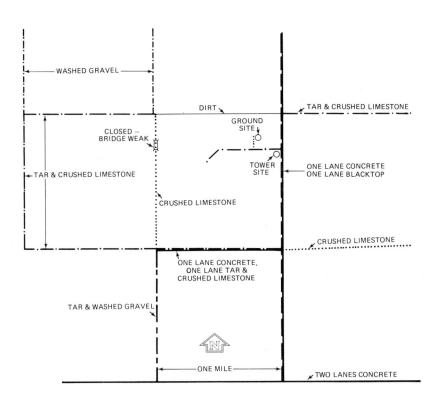

Figure 1. Map of the sampling site and the surrounding area showing ground and tower sampling sites and road surface characteristics. During the growing season the area is planted exclusively in corn and soybeans.

306   ATMOSPHERIC AEROSOL

California-Davis, using ion-excited X-ray fluorescence (3, 4).
Further details of the daily Nuclepore™ sampling and analysis
procedures were given in an earlier publication (5).
    Hourly wind directions were obtained from measurements
made by Federal Aviation Agency personnel at the University
of Illinois Willard Airport, about 8 km east of the sampling
site.

    Source Materials. Two categories of local source materials
were sampled - road dust and soil. Samples of bulk soil were
removed from the upper 10 cm of the soil column at 74 locations
over portions of the 1 km² centered on the ground site. A square
grid spacing of 160 m (0.10 mi) between sampling locations
was used. Before analysis, the samples were air dried,
crushed, and sieved to remove the larger particles (>53 μm
diameter) which would not be likely to be raised by the wind.
    Road materials were gathered from 9 locations on unpaved
roads in the vicinity of the sampling site. At each location,
dust from the roadbed was scraped into a polyethylene sample
bag from 5 or more spots over a 10 m² area. These materials
were also air dried and sieved identically to the soil materials,
but not crushed. Visual inspections of the road surface
suggested that samples from 8 of the 9 sites were primarily
crushed limestone rock. The remaining site consisted primarily
of washed gravel derived from glacial deposits. An inventory
of unpaved road surface materials within 5 km of the sampling
site indicated that approximately 80% consisted of crushed
limestone.
    Analysis of both soil and road dust for Al, Si, K, Ti, and
Fe was performed at the Illinois State Geological Survey,
using X-ray fluorescence techniques (6). Ca and K were determined at the Water Survey by atomic absorption spectrophotometry
(AAS) following LiBO₃ fusion to disolve the sample (7, 8).
The K analyses employed an air/acetylene flame, but it was
found necessary to use N₂O/acetylene for accurate results on Ca.
    A series of four standard soils, which are certified
reference materials from the Canada Centre for Mineral and
Energy Technology, Ottawa, was analyzed by the AAS/fusion
procedure to assure that it gave satisfactory results. Excellent
agreement with recommended values was obtained. Intercalibration
of the two laboratories used for the source materials analyses
was accomplished by the parallel analyses of K. Again,
excellent agreement was achieved.

    Factor Analyses. Analyses were carried out on elemental
concentrations in air (ng m⁻³), or elemental abundances (percent
of total mass). Both kinds of data were standardized by subtracting the element mean from the observed value and dividing
by the element standard deviation. Computations were carried
out using the factor analysis program from the BMDP Biomedical

Computer Programs (9). Initial factors were extracted using a principal components solution. The number of factors to be kept for rotation to a final solution was selected from a plot of the variance explained by each factor (its eigenvalue) versus its ordinal number. Usually, factors with eigenvalues larger than about 1.0 were kept. Final solutions were obtained using Varimax rotations.

Results are presented in terms of tables of "loadings." Loadings are correlation coefficients between variables and factors. Thus, elements with high loadings on the same factor tend to vary in a parallel manner and may often be inferred to have common sources. The sum, across the factors, of the squares of the loadings, is called the communality, and indicates the fraction of an element's total variance that is accounted for by all the factors.

Chemical Element Balances. The method is based on the general relationship

$$P_i = \sum_j p_{ij} C_j \tag{1}$$

which expresses the percentage P_i of any element i in the aerosol as a sum of terms involving p_{ij}, the percent of element i in the particulate matter emitted by source j, and C_j, the fraction of the total aerosol contributed by source j. For additional discussion of the method, see (10).

We assume here that soil and roads are the only non-negligible sources of K and Ca at our rural site, so that

$$P_{Ca} = p_{Ca, soil} C_{soil} + p_{Ca, roads} C_{roads}$$

and

$$P_K = p_{K, soil} C_{soil} + p_{K, roads} C_{roads}$$

We have measured the P's and the p's, so we may solve the simultaneous equations for C_{soil} and C_{roads}. The percent contribution of soil to the airborne Ca is then $100 \, p_{Ca, soil} \, C_{soil}/P_{Ca}$. This method, although simple, is justified in this case because the two sources considered are the only major ones, and because the abundances of Ca and K are similar, requiring no normalization, as is sometimes needed (10).

Once the C's have been computed, they may be used with the measured p's of other elements to compute the P's for those elements. These may by compared to observed values as a check on the results. The observed source compositions (p's) used in the calculations, and the mean elemental abundances (P's) of the 36 aerosol samples for which C's were computed separately, are given in Table 1.

Table I. Mean abundances of crustal elements in aerosols,
soil, and road dust in a rural area near
Champaign, IL.

	P_i %	P_i, soil %	P_i, roads* %
Al	1.8	5.1	1.30
Si	6.7	35	13.1
K	0.78	1.75	0.52
Ca	2.2	0.70	25
Ti	0.24	0.42	0.08
Fe	1.3	2.2	1.29

*Assuming all unpaved road surfaces are crushed limestone

Results

Distribution of Concentrations by Wind Direction. The good
time resolution of the streaker data means that single 2-hr time
segments often represent a very narrow range of wind directions.
This is potentially important for source identification, since
it allows one to examine variations of concentrations as a
function of wind direction. Table II shows overall mean
concentrations of the crustal elements Ca, Al, Si, and K for
four sampling periods in 1978, and Figures 2 to 5 show variations of concentration with wind direction during these periods.
     Figure 2 shows results at the ground site for a one-week
period in June, when crops (corn and soybeans) were short
enough to allow the generation of soil aerosols through
cultivation and wind erosion. Figure 3 shows results from the
tower sampler for a two-week period in July when crops were
mature. Cultivation should have been absent, and wind erosion
very limited, during this period.
     Figure 4 and 5 show results for a period during and after
harvest, in October and November, 1978. Generation of soil
aerosol by tilling was observed, and wind erosion was possible,
during this period because of the removal of the crop cover
from the soil surface. Figure 4 shows results of three weeks
of sample collection from the ground site, and Figure 5 shows
results from four weeks of sampling on the tower.

Table II.   Mean concentrations of four important crustal
elements in four aerosol sample groups.

	No. 2-hr sample periods	Mean Concentrations, ng m^{-3}			
		Ca	Al	Si	K
June, ground	82	1930	780	3600	730
July, tower	169	640	160	810	350
Fall, ground	249	1990	520	1850	620
Fall, tower	332	1800	490	1420	610

Note in Figure 2 that SSE winds produced the maximum Ca
concentrations, with smaller values from the SSW, at ground
level in June, whereas the order was reversed for the other
three crustal elements. This suggests a different primary
source for Ca than for Al, Si, and K during this period. The
limestone gravel road 1 km SE (see Figure 1), which we observed
to produce a dense white cloud of limestone dust from vehicle
traffic in dry weather, is a likely Ca source. The limestone
gravel road to the west is not apparent in Figure 1, perhaps
because of less traffic due to a weak bridge.

The most obvious feature of the July tower data in Figure 3
is the overall drop in the concentrations of all four elements.
This is consistent with the idea that tall crops inhibit wind
erosion or collect eroded material within the crop canopy. It
is also consistent with a decrease in concentration with height,
which would be expected for materials having their source at the
surface. However, the decrease was much more than was observed
between the ground and tower sites in the fall (Figures 4 and 5),
and thus would appear to have an additional cause beside the
elevation difference. In July (Figure 3) the maximum concentrations appeared from the SE (the direction of the nearest
limestone gravel road) for Ca and two of the other three crustal
elements. (K concentrations were quite uniform around the compass.) This suggests that roads may be an important source
of Si and Al, as well as of Ca, when mature crops are limiting
the mechanisms of soil erosion; i.e., wind and tilling.

Figures 4 and 5 show results of sampling over about one
month in the fall: mid-October to mid-November 1978. This
period included harvest activities as well as post-harvest
plowing and other tilling operations. These activities were
visually observed to mobilize a considerable amount of soil
material. This is reflected in the increased concentrations over
those of July (Table II). Moreover, mean concentrations
during the fall period approached, and in one case exceeded,
those of June, when tilling and wind erosion were also active.

Figure 4 shows results for the ground level sampler. The
maximum concentrations of all four crustal elementals occurred

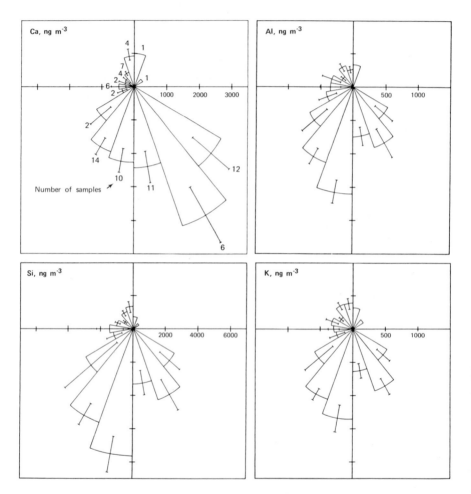

Figure 2.   Distribution of mean concentrations with wind direction for 4 crustal elements measured on a streaker sampler at ground level during one week in June 1978.   Radial bars indicate ±1 standard error of the mean.   The numbers at the end of the bars indicate the number of 2-h samples from each direction.

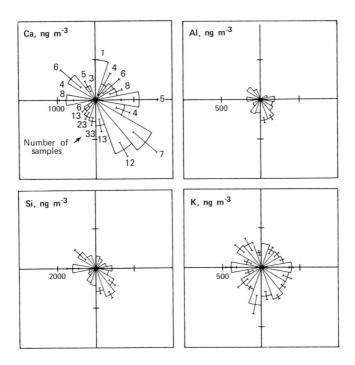

Figure 3. Same as Figure 2, except for a 2-week period on the tower during July 1978.

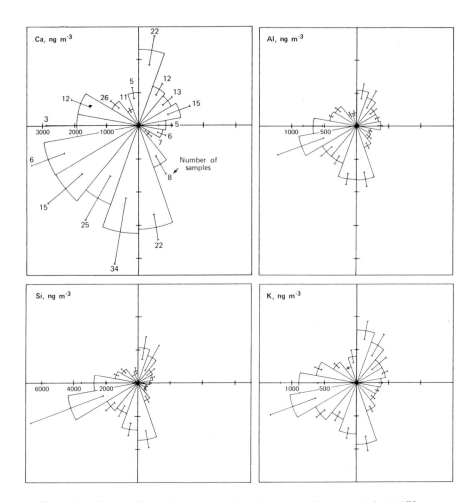

Figure 4. Same as Figure 2, except for 3 weeks at ground level in the fall, 1978.

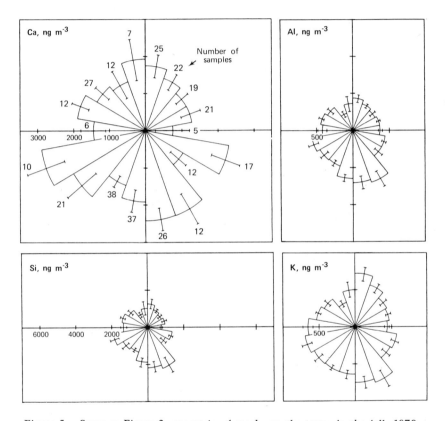

Figure 5. Same as Figure 2, except for 4 weeks on the tower in the fall, 1978.

in winds from SSE through W. For each element there was also a secondary peak from the NNE. The lack of a difference in directional distribution between Ca and the other elements suggests a common primary source for all, and this is consistent with the observations of large amounts of soil and aerosol from agricultural field activity during this period.

Concentrations measured on the tower during the fall period are shown in Figure 5. Compared with concentrations at the ground, these are slightly smaller (see Table II) and more uniform around the compass, but still quite similar element to element. Nothing here suggests different primary sources for these crustal elements, but the seasonal variations in concentration and occasional differences in directional responses of the summer and fall sampling periods suggest that both roads and soil are important sources, at least for Ca.

Abundances. Additional evidence of possible multiple sources comes from an examination of the mass percentages (abundances) of crustal elements in aerosol samples. Figures 6 through 11 show abundances for Al, Si, K, Ca, Ti, and Fe. The figures also show measured abundances in local soil and road dust sources for comparison. Except for Si, the elemental abundances in the aerosol suggest a combination of road and soil sources. The Ca results are important to this observation because only for Ca were the abundances in aerosols consistently less than those in road dust, but greater than those in soil. From the other elements, one could reasonably conclude that roads were the major source.

The Si abundances in the aerosol generally fall below those of both roads and soil, but they are much closer to those of roads. As will be shown later from other evidence, there is reason to believe that Si abundances are diminished in aerosol relative to bulk surface materials. This is consistent with our knowledge that Si is a major component of the larger particles of our soil samples, which are either too big to become airborne, or, if raised by the wind, tend to fall out quickly.

Factor Analyses. In an attempt to identify sources of the various elements, factor analyses were carried out separately on two data sets representing 1) filters exposed face down under an inverted funnel rain shield, and 2) filters exposed in a vane sampler continuously facing into the wind. Within each data set, separate factor analyses were performed on the data expressed as 1) concentrations in air (ng m^{-3}) and 2) abundances (percent of total mass). Rain amount, rain duration, and soil moisture data were included in early analyses, but these parameters were later dropped from the data sets because they had no significant relationship to any of the elements. Wind direction frequencies were included in the data sets throughout the analyses, however.

Different results were obtained for concentration data

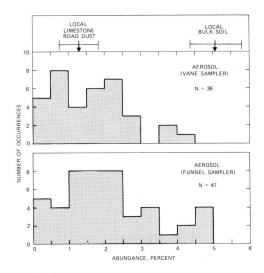

Figure 6. Frequency distributions of Al abundances from 2 sets of aerosol samples, summer, 1978. Source abundances for Al are shown for comparison. The bars on the source abundances indicate ±1 standard deviation of an individual source sample.

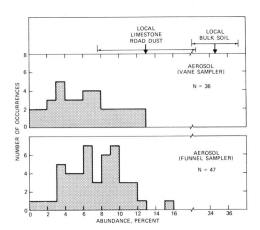

Figure 7. Same as Figure 6, except for Si

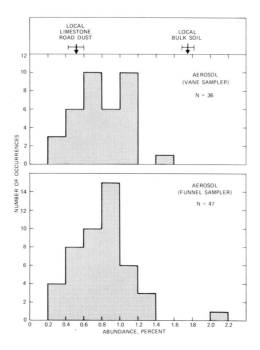

Figure 8. Same as Figure 6, except for K

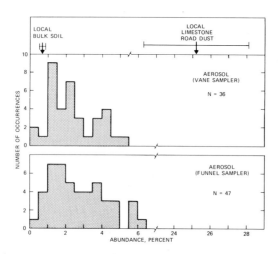

Figure 9. Same as Figure 6, except for Ca

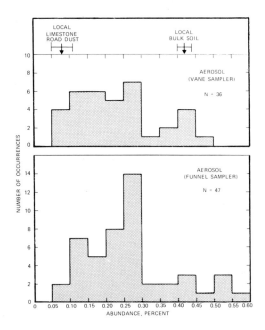

Figure 10.   Same as Figure 6, except for Ti

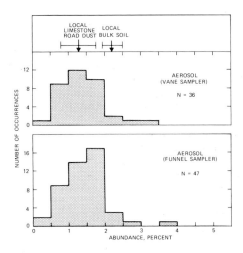

Figure 11.   Same as Figure 6, except for Fe

than for abundance data, but differences between samplers (vane vs. funnel) were minimal. Therefore, the results presented here, for both concentrations and abundances, are for vane samples only.

Table III shows the loadings table based on concentrations. Factor 1, which explains 44% of the variance, includes all the usual crustal elements at high loadings plus Zn and Pb at much weaker loadings. Factor 2 includes Pb and S, but only at relatively weak loadings, associated with SE winds. This direction is consistent with the long range transport of sulfur from the Ohio Valley, but no obvious Pb sources are apparent in the SE direction. Note that Ca is present on only one factor, with a relatively high loading (0.87).

The loadings for the data expressed as abundances are shown in Table IV. Now four factors appear, but only the first two contain elements. Factor 1 is a combination of crustal elements and S, with an inverse relationship between the two. This is reasonable for major components of the mass because when expressed as abundances, any extremely high values of one component automatically decrease the other. As with concentrations, no wind directions are important on factor 1. Factor 2 contains the pollutant metals Pb, Zn, and Mn, as well as Ca and Fe at weak loadings, and tends to occur with NE winds. This suggests pollutant transport from Champaign-Urbana.

In this case we do find Ca split between sources, but not the ones expected. The apparent road source to the SE, suggested in the concentration distributions with wind direction, does not appear in the factor analysis results, no matter whether based on concentrations or abundances. This result contrasts with those presented earlier, which suggested a combination of road and soil sources, at least for Ca.

Chemical Element Balance. To help resolve this dilemma, the results of a simple chemical element balance based on Ca and K are useful. The calculations were performed separately for each of the 36 vane samples collected, so a distribution of the results could be presented. The input data for these calculations included the mean abundances of the elements in the source materials (local bulk soil and road dust) (Table 1) and the aerosol element abundances measured in the 36 individual vane samples. The distributions of crustal element abundances in aerosols were shown in Figures 6-11 along with the respective abundances of the elements in the source materials.

The distributions of the contributions of roads and soil to total suspended particulate matter (TSP) are shown in Figure 12. The figure shows that soil contributed a mean of 39% of the TSP and roads 12% during the summer sampling period in 1978. The distributions show that on individual days soil contributed up to about 70% of the TSP and roads up to about 30%.

Table III. Loadings from factor analysis of vane sampler data, based on concentrations. Loadings <0.50 have been omitted for clarity.

	Factor			
	1	2	3	Communality
Fe	0.94			0.95
K	0.92			0.94
Al	0.92			0.89
Si	0.90			0.95
Ti	0.88			0.94
Mn	0.88			0.80
Ca	0.87			0.80
Zn	0.66			0.48
SE*		0.83		0.70
NW*		-0.78		0.68
Pb	0.56	0.64		0.78
S		0.62		0.47
SW*			0.90	0.84
NE*			-0.77	0.65
V*				0.17
Variance explained, %	44	15	15	
Cumulative variance, %	44	59	74	

*Wind direction categories are northeast, southeast, southwest, northwest, and variable.

Table IV.   Loadings from factor analysis of vane sampler data,
            based on abundances.   Loadings <0.50 have usually
            been omitted for clarity.

	Factor 1	Factor 2	Factor 3	Factor 4	Communality
K	0.90				0.85
Al	0.88				0.77
Si	0.87				0.87
S	-0.86				0.80
Fe	0.79	(0.49)			0.90
Ti	0.73				0.79
Ca	0.69	(0.43)			0.78
Mn		0.88			0.81
Pb		0.85			0.82
Zn		0.76			0.79
NE*		0.77			0.74
SE*			0.89		0.74
NW*			-0.88		0.87
SW*				0.84	0.85
V*				-0.67	0.49
Variance explained, %	32	23	13	12	
Cumulative variance, %	32	55	68	80	

*Wind direction categories are northeast, southeast, southwest,
northwest, and variable.

Other major contributors to TSP (not measured) are sulfates, organic matter, and water.

Source contributions to airborne Ca may be calculated from the contributions to TSP by multiplying the source fractions by their respective Ca contents. Figure 13 shows the results of these calculations in the form of the distribution of the percent airborne Ca contributed by the soil. On the assumption that roads and soils are the only sources of airborne Ca, a mirror image of this distribution would appear on the right side of the figure for the percent airborne Ca contributed by the roads. For the period of sampling we see that roads contributed a mean of 86% of the Ca, and soil 14%. Even allowing for reasonable uncertainty in these figures, roads appear to be the major Ca source in rural central Illinois in the summer.

A check of the results may be made by substituting the calculated Ca and observed source compositions of other elements into Equation (1). The P values computed in this way may then be compared against observed values. Such a comparison appears in Table V, assuming only road and soil sources for the elemented shown.

Table V. Comparison of calculated and observed abundances in aerosol.

	P_i	
	calculated	observed
Al	2.1	1.8
Si	15.2	6.7
Ti	0.17	0.24
Fe	1.0	1.3

The calculated abundances for Al and Si were higher than actually observed. The Al value is only slightly higher than observed, and probably within the expected error. Si, however, is more than double the observed value, and appears to be evidence of element fractionation in the process of soil aerosol generation. This is reasonable in the case of Si because it is known to be a major component of the larger particles of our soil samples which should become airborne only in unusual situations, and which would rather quickly fall out of the atmosphere. Ti and Fe were observed to be more abundant than expected from roads and soil alone, and may have additional sources. For Fe, this is consistent with the factor analysis results which showed a weak Fe loading on the "urban pollution" factor (Factor 2, Table IV) that appeared when using abundances as input data.

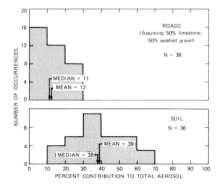

*Figure 12.   Frequency distribution of the
percent contribution of soil and roads to
TSP (vane sampler—summer 1978)*

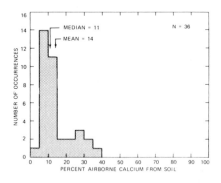

*Figure 13.   Frequency distribution of the
percent contribution of soil to airborne
Ca (vane sampler—summer 1978)*

Discussion

The results presented a variety of evidence for the identity of Ca sources near our rural sampling site. The distribution of mean crustal element concentrations as a function of wind direction in summer and fall, from the streaker data, suggest a combination of road and soil sources. This agrees with a comparison of crustal abundances in aerosols and source materials. The comparison showed that most of the elements examined had abundances in the aerosol that often fell between those characteristic of roads and soil. This was not the case for Si, but Si may be expected to be less abundant in aerosol samples than in bulk surficial materials because of the preponderance of quartz (SiO_2) in the larger particles.

The indications of multiple sources for Ca was not confirmed by the factor analysis procedure used, although the data were analyzed as both concentrations and abundances. Ca was loaded most heavily on factor 1 in all analyses, accompanied by all of the other crustal elements, such as Al, Si, K, Ti, and Fe. When the data were factor analyzed in abundance form, a weak Ca loading appeared on a factor that appears to be composed of urban pollutants from Champaign–Urbana, including Pb and Zn. This is of considerable interest, because urban concentrations of Ca and other crustal elements have been found to be enhanced over rural concentrations (11, 12, 13). Nevertheless, the factor analysis procedure used did not distinguish between rural soil and road sources of crustal elements.

However, the results of the chemical element balance, based on assumed road and soil sources, were consistent with the multiple sources implied by the observed abundances, the seasonal variations in concentration, and the wind direction dependence. The chemical element balance shows that in summer 1978 roads contributed a mean of 12% to the local TSP and soils a mean of 39%. If these were the only Ca sources, this means that roads contributed 86% and soil 14% of the airborne Ca.

The relative contribution of roads and soils to TSP and to airborne Ca in other rural areas will vary with the composition of rock materials used on unpaved roads, with traffic density, and with the distance of the receptor from upwind roads. We have not carried out an extensive analysis of the effects of these variables. However, the distributions presented in Figures 12 and 13 show the range of results obtained and the sensitivity of the method to variations of Ca abundance in aerosol by using actual data from 36 days during a single summer. Further, a separate calculation using the Ca abundance measured in one sample of washed gravel road dust, and assuming 50 washed gravel and 50% limestone roads changed the percent Ca due to soil by only 2%.

These results strongly indicate that roads are an important source of airborne alkaline materials in the summer, when the soil contribution is expected to be suppressed.

Conclusions

Airborne Ca at a rural site in central Illinois comes from
both roads and soil in the summer. If these are the major
sources, which appears to be the case, the road contribution
averaged about 86% and the soil contribution 14% in the summer,
1978. However, roads contributed a mean of only 12% to the TSP,
and soil 39%.
The factor analysis technique used was unable to distinguish
separate soil and road sources. Ca appeared with Al, Si, K, Ti,
and Fe on a factor that can be characterized only as "crustal,"
including both soil and road materials. It appears that a chemical element balance should always be used as a check on factor
analysis results, at least until a more sophisticated factor
analysis method, such as target transformation factor analysis
(14), can be shown not to require it.

Acknowledgments

Sample analyses were carried out by a number of laboratories.
We are grateful to Mr. Mark E. Peden and Ms. Loretta M. Skowron
of the Water Survey's Analytical Chemistry Laboratory Unit for
atomic absorption spectrophotometry, Mr. L. R. Henderson of the
Illinois State Geological Survey for X-ray Fluorescence specto-
scopy, and Dr. T. A. Cahill of the University of California-Davis
for elemental analysis. Mr. R. G. Semonin reviewed the manu-
script. This material is based upon work supported by the
National Science Foundation under Grant No. ATM-7724294, and by
the Department of Energy, Division of Biomedical and Environmental
Research, under Contract No. EY-76-S-02-1199.

Literature Cited

1. Nelson, J. W., Proton-induced aerosol analysis: methods and
    samplers. In Dzubay, T. G., Ed., "X-Ray Fluorescence of
    Environmental Samples," Ann Arbor Science Publishers, Ann
    Arbor, Michigan, 1977, 19-32.
2. Johansson, T. B., Van Grieken, R. E., Nelson, J. W., and
    Winchester, J. W., Elemental trace analysis of small samples
    by proton induced X-ray emission. Anal. Chem., 1975, 47,
    855-860.
3. Flocchini, R. G., Feeney, P. J., Sommerville, R. J., and
    Cahill, T. A., Sensitivity versus target backings for
    elemental analysis by alpha-excited X-ray emission. Nucl.
    Insts. and Methods, 1972, 100, 379-402.
4. Flocchini, R. G., Cahill, T. A., Shadoan, D. J., Lange, S. J.,
    Eldred, R. A., Feeney, P. J., and Wolfe, G. W., Monitoring
    California's aerosols by size and elemental composition.
    Environ. Sci. & Technol., 1976, 10, 76-82.

5. Gatz, D. F., Indentification of aerosol sources in the St. Louis area using factor analysis. J. Appl. Meteorol., 1978, 17, 600-608

6. Rose, H. J., and Flanagan, F. J., X-ray fluorescence determination of thallium in manganese ores. "U. S. Geological Survey Professional Paper 450-B," 1962, 80-82.

7. Van Loon, J. C., and Parissis, C. M., Scheme of silicate analysis based on the lithium metaborate fusion followed by atomic absorption spectrophotometry. The Analyst, 1969, 94, 1057-1062.

8. Suhr, N. R., and Ingamells, C. O., Solution technique for analysis of silicates. Anal. Chem., 1966, 38(6), 730-734.

9. Dixon, W. J., "BMDP Biomedical Computer Programs," University of California Press, Berkeley, 1975, 792 pp.

10. Kowalczyk, G. S., Choquette, C. E., and Gordon, G. E., Chemical element balances and identification of air pollution sources in Washington, D.C. Atmos. Environ., 1978, 12, 1143-1153.

11. Neustadter, H. E., Fordyce, J. S., King, R. B., Elemental composition of airborne particulates and source identification: data analysis techniques. J. Air Poll. Control Assoc., 1976, 26(11), 1079-1084.

12. Hardy, K. A., Akselsson, R., Nelson, J. W., and Winchester, J. W., Elemental constituents of Miami aerosol as a function of particle size. Environ. Sci. & Technol., 1976, 10(2), 176-182.

13. Gatz, D. F., Airborne elements: their concentration distributions at St. Louis. In Semonin, R. G., Ed. "Study of Air Pollution Scavenging," Fifteenth Progress Report, Contract No. EY-76-S-02-1199, U. S. Department of Energy, Illinois State Water Survey, Urbana, 1977, 26-50.

14. Alpert, D. J., and Hopke, P. K., A quantitative determination of sources in Boston urban aerosol. Atmos. Environ., 1980, 14, 1137-1146.

RECEIVED March 10, 1981.

The Effect of Owens Dry Lake on Air Quality in the Owens Valley with Implications for the Mono Lake Area

J. B. BARONE, L. L. ASHBAUGH, B. H. KUSKO, and T. A. CAHILL

Air Quality Group, Crocker Nuclear Laboratory and Department of Physics, University of California—Davis, Davis, CA 95616

The suspension of aerosols during dust storms from the Owens Dry Lake in California has been a subject of great concern to residents in the Owens Valley. In order to assess the magnitude and extent of aerosols from the lake bed, we measured ambient aerosol concentrations by size, mass, and elemental concentrations at seven sites in the Owens Valley from February 20 to June 18, 1979. Resuspended samples of material from Owens Lake bed were used to determine the lake bed source signature. The most important effect we measured was a significant increase in sulfate aerosols as far as twenty-five miles downwind of the lake bed. A substantial increase in gravimetric mass was measured during dust storms for particles in the 2.5-15µm size range. No significant increase in mass was measured for particles in the 0.1-2.5µm size range. Similar measurements on a more limited scale at Mono Lake showed that the sulfur to iron ratio at Mono Lake is a factor of 6-8 greater than the ratio at Owens Lake. These results are consistent with the known mechanism of sulfate formation (efflorescence) in alkaline lake beds. The data indicate that Mono Lake has the potential to become a significant source of sulfate aerosols if the lake bed areas are exposed.

The Owens Valley lies in east central California between the Sierra Nevadas to the west and the Inyo-White Mountains to the east. The valley is 120 miles long, but only seventeen miles across at its widest point. While the valley floor lies at an elevation of approximately 4,000 feet, the mountains on either side tower to 12-14,000 feet. This topography is important for two reasons. First, the Sierra Nevadas create a rain shadow

commonly associated with the development of saline lakes (1). In
fact, several saline lakes and playas exist in the vicinity, including Owens Dry Lake, Mono Lake, Saline Valley, Deep Springs
Lake, Death Valley, and Searles Lake. Second, the high mountains
on either side constrain the transport of airborne contaminants
within the valley, effectively trapping them until they exit at
the north or south end. Furthermore, the mountains may serve to
increase wind speeds at the valley surface due to local chinooks
and venturi flow effects between the valley walls (2). This effect could enhance development of dust storms in the valley (3).

In recent years, the importance of the Owens Lake bed on
air quality in the Owens Valley has been a subject of much concern and debate. The extent to which the lake bed aerosols are
transported, as well as whether or not toxic substances are present in the lake bed soils, has been questioned by residents and
regulatory agencies. Although air quality in the basin is generally quite good, at some times during the year, particulate aerosol concentrations are high. In particular, dust storms occur
during windy periods, which cause poor visibility and may lead to
health problems (2).

In order to obtain quantitative data on particulate air
quality in the Owens Valley, a study sponsored by the California
Air Resources Board was conducted by the Air Quality Group at
UCD. The primary objective was to determine the impact of the
dry lake bed on the average particulate concentration and on the
dust storm particulate concentrations in the valley. In order to
accomplish this, it was necessary to determine the elemental composition of the dry lake bed and to determine the average weekly
and dust storm concentration of aerosols.

The objectives were met by a study plan designed to acquire
data for seventeen weeks at seven sites throughout the valley.
This included both weekly monitoring and daily intensive sampling
of aerosols. The samples were weighed and then analyzed for elemental composition on the UC Davis cyclotron.

It should be noted that the sampler used in the study, (a
stacked filter unit), collected only particles less than 15μm
aerodynamic diameter, i.e. particles of respirable size. State
standards for TSP are based on High Volume samplers which have no
inlet cutoff. Hence, particles as large as 100 microns can be
captured by these instruments. Therefore, measurements made in
this study may not indicate whether particulate standards have
been violated, since a significant portion of the total suspended
particulate mass is not measured by the stacked filter unit (SFU).

Particle Sampling and Analysis

The particle sampler chosen for this study was the Stacked
Filter Unit (SFU) described by Cahill et al (4,5). Particle collection in two size fractions was achieved by placing two Nuclepore membrane filters in series. The first filter, with 8μm

diameter pores, has a 50% collection efficiency for 2.5μm diameter particles. The filters were coated with a thin layer of Apiezon type L grease to minimize particle bounce-off and loss in transport. The SFU was fitted with an inlet which excluded particles larger than 15μm. The second filter was a 0.4μm pore diameter Nuclepore filter, which collected all particles less than 2.5μm. A diagram of the SFU is shown in Figure 1, and the collection efficiency of each stage is shown in Figure 2. Also shown is the bimodal "Pasadena" particle size distribution (6), and the collection efficiency of the human upper respiratory tract (7). The SFU thus collects particles in two size fractions closely resembling the fractions deposited in the upper and lower respiratory tracts, and also separates the coarse, naturally generated particles from the finer anthropogenic particles.

The aerosol samples collected by the SFU were analyzed both gravimetrically for total suspended particulate mass less than 15μm, and by particle induced x-ray emission (PIXE) for elemental content. The filters were weighed before and after sampling using a Cahn 25 electrobalance sensitive to 1μg. Typical precision of TSP determined by this analytical method is ±0.5μg/m^3 for samples collected under conditions of low aerosol concentrations (5). After weighing, the filters were analyzed for elemental content (elements heavier than Na) using the UC Davis PIXE system. This analysis technique is described in Cahill et al (8).

Study Design

Weekly monitoring of particulate aerosols began on February 20, 1979, and ended on June 18, 1979. A total of seventeen weeks of monitoring were conducted during this period. Samplers were run for seven consecutive days each week at a flow rate of ten liters per minute, except during dust storm episodes. During dust storms, samples were collected daily at a flow rate of ten liters per minute. Samples during dust storms were collected on April 6,7,16,17,23,24, 1979.

In addition, data on wind flow collected at Bishop was examined. This station measured wind speed and direction between 7 AM and 7 PM. However, these data may not be representative of the flow regime at the Lake since they do not include a complete 24-hour record and are a significant distance from the dry lake.

The sampling sites were chosen in order to investigate the spatial distribution of particulate pollutants in the valley. As an additional consideration, sites were selected to coincide with the major population centers in the valley in order to determine the concentration of respirable aerosols to which valley residents are exposed on a daily basis. Seven of the sampling sites were in the Owens Valley itself, and one site was in the Mono Lake area. Site 1 was located near the Bishop Airport at the National Weather Service Meteorological station. This site is about five miles east of downtown Bishop in the center of the

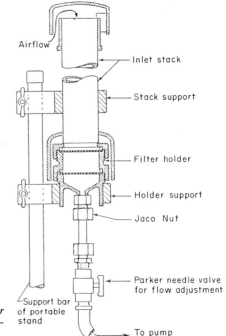

Figure 1. Schematic of the stacked filter unit showing the inlet and the 2-stage filter holder

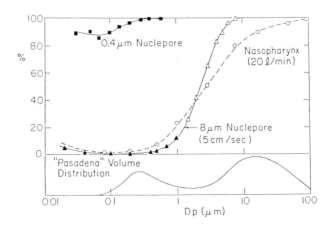

Figure 2. Particle capture efficiencies of 8.0-μm and 0.4-μm Nuclepore filters for a face velocity of 5 cm/s.

Also shown is the capture efficiency of the human upper respiratory tract (nasopharynx). Closed symbols are derived from data on capture efficiency using monodispersed aerosols; open symbols are theoretical calculations (see Ref. 4). The lower section of the diagram is a schematic of the "Pasadena" bimodal distribution.

valley. Site 2 was located in Big Pine near the western edge of
the valley. The Independence courthouse was the location of Site
3. This site was in the center of Independence near Highway 395.
Site 4 was located at the Lone Pine High School on the southeast
side of town. This site is about ten miles north of Owens Dry
Lake. The Keeler Post Office was chosen for Site 5. This site
was on the eastern edge of Owens Dry Lake. Site 6 was located at
Cartago on the western side of Owens Dry Lake. Site 7 was located approximately twenty-five miles south of Owens Dry Lake at
Little Lake. Figure 3 is a map of the study area depicting the
seven sampling sites in the valley. Site 8 was located at Lee
Vining in the Mono Lake area on the northwestern edge of Mono
Lake.

All eight sampling sites were operated by local residents.
Preweighed filters were placed in filter holders at U.C. Davis
and shipped via U.P.S. to each site. The local operator would
measure the flow before and after sampling with a spirometer calibrated orifice meter, and then return this information with the
exposed filters. Upon arrival at U.C.D., filters were post-
weighed and prepared for x-ray analysis.

Results

Average Weekly Concentration of Aerosols. The weekly monitoring data colleted in this study provided an excellent means by
which the profile of pollutants in the valley could be determined.
Figure 4 is a profile of the average total and fine gravimetric
mass measured at each site in the basin. The error bars represent the standard error of the mean. A map of the study area is
included at the bottom of the figure to provide a better understanding of the relationship between topography and the measured
pollutant concentrations. The dominant characteristic of this
profile is the marked peak in total mass at the Keeler sampling
site. This peak, coupled with the decreasing total mass values
measured farther north from the dry lake bed, suggests that the
dry lake bed is a significant source of coarse aerosols. The
fine aerosol mass does not follow the same pattern as the total
mass, suggesting that the lake bed may not be a significant
source of fine aerosols. A similar profile depicting selected
trace elements is shown in Figure 5. The error bars here are
again the standard error of the mean. Silicon and iron are usually soil-derived aerosols, while sulfur is generally produced by
gas to particle conversion of sulfur dioxide to sulfate aerosols
or by direct production of sulfates through the burning of fossil
fuels. The source of iron and silicon in the Owens Valley appears to be local soil material with only slight enhancement due
to the dry lake bed. There is no anthropogenic source of coarse
sulfur aerosols near the dry lake. Thus, these aerosols are being suspended from the dry lake bed.

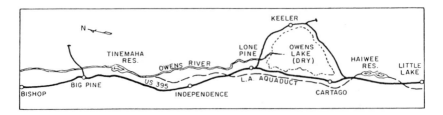

Figure 3.   Map of the Owens Valley with the aerosol sampling sites indicated by
(○)

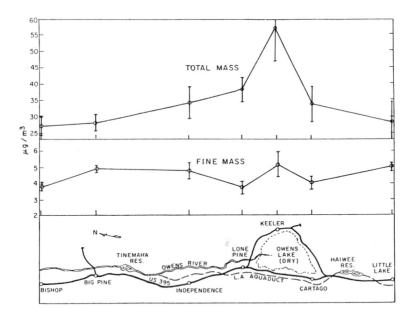

Figure 4.   Average for all weeks of aerosol mass less than 15.0 μm (total mass)
and mass less than 2.5 μm (fine mass) at each sampling site.   A map of the sampling
sites is included at the bottom of the figure.   Error bars represent standard error of
the mean.

Effect of Dust Storm Episodes on the Average Weekly Aerosol
Concentrations. The total and fine gravimetric mass averaged
over all sites for each week, is depicted in Figure 6. The error
bars for the Owens Valley curves represent the standard deviation
of the mean. The errors on the Mono Lake curve represent the
sampling system error of ±15%. The mean weekly values do not include the three dust storm episodes sampled separately, but do
include several additional dust storms. Table I lists all the
dust storms reported by the sampler operators.

The Fine Gravimetric Mass is virtually unchanged over the
monitoring period. The Total Gravimetric Mass (TGM) measurements,
however, illustrate several important points. First, a general
upward trend from winter to summer is observed. This trend has
been observed in data from other sites in Utah and can be attributed to higher levels of soil aerosols suspended in the atmosphere as the soil surface dries out. Superimposed on the general trend, however, are peaks of TGM which are a factor of 1.3 to
2.3 times the non-peak levels. Each of the peaks corresponds to
a dust storm or high winds period as shown in Table I. Furthermore, the dust storm episodes which were sampled separately in
the first, third, and fourth weeks of April, would raise the TGM
in those periods by a factor of 1.7. It should be noted that the
third week of April already includes one dust storm period which
was not sampled separately. Finally, although it is difficult to
draw inferences from these data about the persistence of suspended dust after the storm, the frequency of occurrence of elevated
TGM levels is well documented. During the seventeen week monitoring period, nine weeks exhibit elevated TGM levels if the separate dust storm samples are included in the weekly average values. Hence, these data indicate that dust storm episodes have a
significant impact on the average aerosol concentration in the
Owens Valley.

Magnitude and Spatial Extent of Lake Bed Aerosols. The magnitude and spatial extent of the contribution of lake bed aerosols to the measured atmospheric aerosol concentration in the
valley can be identified using data from the intensive dust
storm sampling program, the weekly monitoring study, and the lake
bed soil sample analysis. Gravimetric mass and four elements,
sulfur, chlorine, silicon, and iron, were selected for analysis.
Based upon data obtained by resuspending lake bed and sampling
site soil samples, two elements (sulfur and chlorine) were identified as being primarily generated in the valley from resuspended lake bed materials. The coarse aerosol fraction (15μm to
2.5μm) of these elements is a good tracer of lake bed materials
for three reasons. First, since these elements were only found
in lake bed soil samples, no other "natural" source of these aerosols is likely to exist in the valley. Second, since the only
anthropogenic source of sulfur aerosols in the valley (automotive
exhaust) would produce fine particle sulfur, (<2.5μm), and since

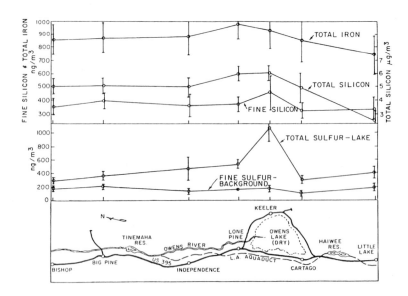

Figure 5.   *Average weekly concentration at each sampling site for total Fe, Si, and S (particles less than 15 μm) and fine Si and S (particles less than 2.5 μm). Error bars represent standard error of the mean.*

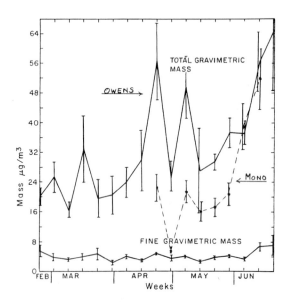

Figure 6.   *Average total mass (particles less than 15 μm) and fine mass (particles less than 2.5 μm) for all Owens Valley sampling sites for each sampling week.*

Error bars are standard error of the mean. Also included for reference is the weekly mass concentration measured at the Mono Lake sampling site. Error bars are ±15% measurement system error.

TABLE I. WEATHER OBSERVATIONS FROM OPERATORS

DATE	MONITORING PERIOD AFFECTED (Month,Week)	OBSERVATIONS NOTED BY SAMPLER OPERATOR
2/26 - 2/21	February, 1	Rain, snow at Big Pine
3/01 - 3/02	March, 1	Dust, North wind
3/15	March, 3	Dust, South wind
3/18, 3/19	March, 3	Heavy rain
4/07 - 4/08	*	Dust storm
4/16	April, 2	High wind at Lone Pine
4/17 - 4/18	*	Dust storm
4/21 - 4/22	April, 3	Big dust storm
4/24	*	Dust storm
5/05	May, 1	Dust, high winds
5/26	May, 4	Dust at Lone Pine
6/15 - 6/18	June, 3	High winds at Big Pine

* These dust storm aerosol measurements are not included in the weekly average concentration in Figure 7.

there is no anthropogenic source of chlorine in the valley, man-
made emissions of coarse particle sulfur or chlorine is negligible. Finally, since long-range transport of coarse particles is
unlikely due to the short residence time of these large particles
in the atmosphere, long-range transport from outside the basin is
not an important source of coarse particle sulfur or chlorine.

Silicon and iron are generally considered to be tracers of
resuspended soils. For this reason, they were included in the
analysis to determine the importance of local soils on the measured aerosol concentration.

Figures 4 and 7 indicate the spatial profile of total and
fine aerosol mass for the weekly monitoring and the dust storm
episode studies, respectively. The total mass concentration
peaks at Keeler and declines sharply north of the lake bed.
These figures indicate that lake bed aerosols are transported in
significant concentrations to Independence. A similar trend is
shown in Figure 8 for total sulfur and coarse silicon and iron
concentrations during a dust storm. The error bars on Figure 7
and 8 represent the sampling system error of ±15%. Weekly monitoring profiles of total silicon and iron concentrations do not
exhibit the same strong peak at Keeler (see Figure 5). These data indicate that coarse sulfur is a good tracer of the aerosols
suspended from the lake bed. Fine sulfur aerosols do not follow
the same trend as coarse aerosols and are probably due to gas-to-
particle conversion processes in the valley and long range aerosol transport into the valley. The weekly monitoring profile of
fine aerosol mass also indicates that fine particles are not generally produced by suspension of lake bed aerosols. The dust
storm profiles, however, do indicate that an increase in fine
aerosol mass occurs near the lake bed, suggesting that during
dust storms, the dry lake is a significant contributor to the
fine atmospheric aerosol concentrations. This may be due to the
high winds which occur during dust storms. During these high
wind periods, saltation and creep of intermediate particles may
be increased significantly (9). This action would aid in the
suspension of fine particles from the surface of the dry lake.
Hence, increased fine particle concentrations near the lake bed
would be measured.

As an indication of the effect of dust storm episodes on the
aerosol concentration in the valley, the per cent increase in the
weekly total mass, coarse sulfur, chlorine, silicon, and iron
concentration during a dust storm was computed. In addition, the
absolute increase in these quantities was also computed. The results of this analysis are shown in Table II. These data also
indicate that a significant increase in aerosol concentration due
to suspended lake bed materials occurs as far downwind as Independence. In order to quantify this effect, the sulfur to iron
(S/Fe) and chlorine to iron (Cl/Fe) ratio at each site was examined. At Keeler, all the coarse sulfur and iron measured at the
sampling site are suspended from the lake bed. At any site

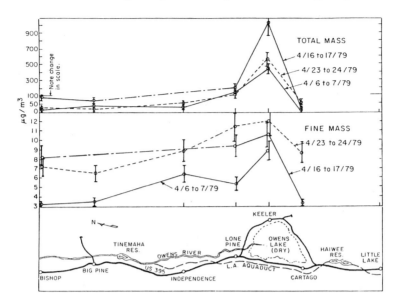

Figure 7.   Total mass (particles less than 15 μm) and fine mass (particles less than 2.5 μm) measured during 3 dust storms: April 6–7, April 16–17, and April 23–24, 1979. Error bars are ±15% measurement system error.

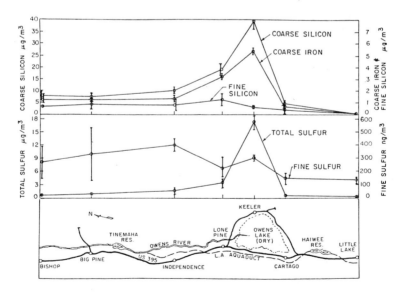

Figure 8.   Mean concentrations of coarse and fine Si, coarse Fe, and total and fine S measured during the 3 dust storm periods. Error bars are standard error of the mean.

TABLE II. PER CENT AND MASS INCREASE OF AEROSOLS DURING DUST STORMS

	BISHOP	BIG PINE	INDEPENDENCE	LONE PINE	KEELER	CARTAGO	LITTLE LAKE
Per Cent Increase							
Total Mass	80 ± 86	45 ± 59	56 ± 22	356 ± 66	1121 ± 384	-14 ± 47	-88 ± 29
Sulfur	3 ± 41	11 ± 57	138 ± 92	631 ± 212	1598 ± 342	137 ± 75	-38 ± 23
Chlorine	87 ± 48	150 ± 135	172 ± 67	1024 ± 442	933 ± 442	367 ± 272	-3 ± 42
Silicon	58 ± 54	27 ± 31	40 ± 29	213 ± 47	556 ± 101	-15 ± 34	-86 ± 34
Iron	61 ± 70	40 ± 48	27 ± 30	186 ± 32	477 ± 72	-23 ± 31	-86 ± 26
Mass Increase							
Total Mass $\mu g/m^3$	22 ± 23	13 ± 17	19 ± 7	137 ± 22	644 ± 186	-5 ± 16	-25 ± 6
Sulfur ng/m^3	11 ± 135	42 ± 216	908 ± 561	3087 ±9115	16111 ±1746	404 ± 209	160 ± 91
Chlorine ng/m^3	61 ± 29	141 ± 120	549 ± 125	2641 ± 973	11414 ±4396	639 ± 430	-7 ±107
Silicon ng/m^3	2977 ±2759	1354 ±1564	2023 ±1455	12806 ±2470	33919 ±2906	-775 ±1704	-3091 ±946
Iron ng/m^3	495 ± 562	343 ± 412	243 ± 271	1810 ± 230	4501 ± 306	-197 ± 262	-652 ±152

downwind from Keeler, all the coarse sulfur and iron measured at the sampling site are suspended from the lake bed. At any site downwind from Keeler, all the coarse sulfur and chlorine is from the lake, but the measured coarse iron and other soil-derived materials are only partly due to lake bed suspended materials. Hence, if we knew the ratio of the coarse iron from the lake to the total iron measured at the site, then we could determine the fraction of the measured aerosol contributed from the lake bed. The following relationship allows us to calculate the ratio of lake bed iron (Fe_L) to total iron measured at a site (Fe_T).

$$\frac{Fe_L}{Fe_T} = \frac{Fe_L/S_L}{Fe_T/S_L} \quad but \quad \frac{Fe_L}{S_L} = \frac{Fe_K}{S_K} \quad where \quad \frac{Fe_K}{S_K}$$

is the ratio of iron to sulfur at Keeler. Furthermore, if all the sulfur at the site if from the lake bed, then

$$S_L = S_T$$

at the site and:

$$\frac{Fe_L}{Fe_T} = \frac{Fe_K/S_K}{Fe_T/S_T} = \frac{S_T/Fe_T}{S_K/Fe_K}$$

The same argument can be made for the chlorine to iron ratio. Hence, by calculating the S/Fe and Cl/Fe ratio at each site and normalizing it to the S/Fe and Cl/Fe ratio at Keeler, it is possible to determine the fractional contribution of elemental lake bed materials to the measured aerosol concentration at the site under study. Furthermore, since the ratio of iron to total gravimetric mass (Fe/Mass) is nearly constant across all seven study sites, it is possible to use the normalized S/Fe ratios as a measure of the total mass contributed by the lake bed at each site. Table III shows the S/Fe and Cl/Fe ratios normalized to Keeler. Also included is the Fe/Mass ratio. The two normalized ratios (S/Fe and Cl/Fe) provide a range of values over which the fraction of aerosols produced from the dry lake bed at each site can be determined. Table IV shows the range of contributions from the dry lake bed to the total suspended particulate (TSP) load at each site. The average weekly aerosol TSP value and the normalized S/Fe and Cl/Fe ratio was used to calculate the values in Table IV. These data also indicate that a significant increase in aerosol mass due to materials contributed by the dry lake occurs as far north as Independence.

The use of Cl and S as lake bed tracers, and Fe as a tracer of non-lake bed soils, is given support by analysis of Owens Lake brine (10). This analysis, as well as analyses of several other

TABLE 3. SULFUR AND CHLORINE TO IRON RATIOS

	BISHOP	BIG PINE	INDEPENDENCE	LONE PINE	KEELER	CARTAGO	LITTLE LAKE
(S/Fe) site / (S/Fe) Keeler	.19 ±.06	.24 ±.08	.56 ±.19	.40 ±.11	1.00 ± .27	.18 ±.04	.33 ±.08
(Cl/Fe) site / (Cl/Fe) Keeler	.10 ±.03	.12 ±.04	.31 ±.10	.29 ±.08	1.00 ± .23	.15 ±.03	.29 ±.07
S/Fe	.22 ±.06	.28 ±.08	.65 ±.19	.46 ±.10	1.17 ± .22	.21 ±.03	.38 ±.06
Cl/Fe	.12 ±.03	.14 ±.04	.37 ±.10	.35 ±.08	1.19 ± .19	.18 ±.03	.35 ±.07
Fe/Mass	.030 ±.005	.030 ±.005	.027 ±.006	.025 ±.006	.016 ± .004	.025 ±.008	.027 ±.004

lakes compiled by Eugster (1) are presented in Table V. The total dissolved solids level is given in ppm, while the individual
constituents of the brine are normalized to the total. The sulfate and chloride ions comprise 60% of the anions present, suggesting that they are appropriate lake bed tracers. Iron is absent as a brine constituent, indicating that it is present only
in the surrounding soils and in the bottom muds of the lake.

TABLE 4. MASS CONTRIBUTION OF THE DRY LAKE BED TO THE
AVERAGE WEEKLY T.S.P. VALUE AT EACH SITE
(Particles Less Than 15μm

SITE	RANGE
Bishop	2.8 - 5.2
Big Pine	3.4 - 6.9
Independence	10.5 - 18.7
Lone Pine	11.3 - 15.1
Keeler	- -
Cartago	5.2 - 6.2
Little Lake	8.2 - 9.1

Hazardous Materials

A detailed analysis of two possible hazardous materials,
sulfur and lead, was made in this study. Figure 6 shows the average weekly spatial profile of sulfur, and Figure 8 shows the
sulfur profile during a dust storm. The error bars on Figure 8
represent the standard error of the mean. Both of these figures
indicate that a significant enhancement of the coarse (15μm-2.5μm)
sulfur mode occurs near the lake bed. The chemical states of the
airborne sulfur can be inferred from information on the geochemistry of the alkaline lake bed. The exposed lake bed consists of
a saline crust overlying alkaline muds. A powdery deposit often
forms on the top of the saline crust when moisture is near the
surface through the mechanism of capillary efflorescence. These
efflorescent crusts have a composition quite different from the
underlying saline crust and mud. They are rich in thenardite
(Na_2SO_4), burkeite ($Na_6(SO_4)CO_3$), trona ($Na_3H(CO_3)_2 \cdot 2H_2O$),
pirssonite ($Na_2Ca(CO_3)_2 \cdot 2H_2O$), thermonatrite ($Na_2CO_3 \cdot H_2O$),
halite (NaCl), and sylvite (KCl). The proportion of sulfur in
these minerals is much higher than in the other components of the
lake bed, and the chemical form of the sulfur is sulfate. The
data on ambient aerosols indicate that these efflorescent crusts
are important in the generation of airborne particles. The sodium to sulfur (Na/S) ratio in bulk samples (mud and saline crust
and efflorescent crust) of the Owens lake bed, resuspended by air
streams in the laboratory, is approximately 33, while the Na/S
ratio in ambient aerosols measured at Keeler is 4 ± 1.

TABLE V. ANALYSIS OF BRINES FROM SELECTED SALINE LAKES

Total in ppm, constituents normalized to total

	OWENS LAKE	MONO LAKE	DEEP SPRINGS LAKE	SEARLES LAKE	SALINE VALLEY
SiO_2	1.4×10^{-3}	2.5×10^{-4}	—	—	1.3×10^{-4}
Ca	2.0×10^{-4}	7.9×10^{-5}	9.2×10^{-6}	4.7×10^{-5}	1.0×10^{-3}
Mg	9.8×10^{-5}	6.0×10^{-4}	3.6×10^{-6}	—	2.0×10^{-3}
Na	.38	.38	.33	.33	.36
K	.016	.021	.058	.077	.017
HCO_3	.24	.096	.028	—	2.2×10^{-3}
CO_3		.18	.066	.081	—
SO_4	.10	.13	.17	.14	.081
CL	.25	.24	.36	.36	.53
TOTAL	213,700	56,600	335,000	336,000	282,360

The Owens Lake brine analysis of Table V indicates that the
Na/S ratio should be approximately 3.8 for lake bed materials,
which agrees quite well with the ambient ratio measured at Keeler.
The above data suggests that airborne sulfur aerosols measured in
the Owens Valley are in the form of sulfates which are suspended
from the efflorescent crust on the Owens Lake bed. Therefore, if
we assume that all the sulfur measured at each site is in the
form of sulfate, then during a dust storm, the sulfate standard
for the state of California ($25\mu g/m^3$) is violated near the Owens
Lake. It should be noted that the sulfate standard was developed
for very fine acidic aerosols. The sulfates measured here are
larger and basic particles, so their toxicity may be different
from particles for which the standard was written. The calculated sulfate levels at each site during a dust storm are listed
in Table VI.

TABLE VI.   TOTAL SULFATE CONCENTRATIONS DURING A DUST STORM
Micrograms Per Cubic Meter

Bishop	1.0 ± 0.4
Big Pine	1.3 ± 0.7
Independence	4.7 ± 1.3
Lone Pine	10.7 ± 2.8
Keeler	51.3 ± 5.2
Cartago	2.1 ± 0.6
Little Lake	0.8 ± 0.1

The weekly average and dust storm profiles of lead are shown
in Figure 9. The error bars represent the standard error of the
mean. Although most lead in the valley is in the fine mode
($<2.5\mu m$), some coarse lead is present at a few sites in the valley. The relatively flat lead profile suggests that this material is produced from automobiles driven in the basin. The increased total lead concentration near Keeler during a dust storm
suggests that the dry lake bed may be a source of these aerosols.
However, the magnitude and spatial extent of the total lead increase during a storm are very small and suggest that lead aerosol suspension from the dry lake bed is not a significant problem.

Mono Lake Monitoring

During the last seven weeks of the Owens Valley study, (April
19 to June 11, 1979), weekly aerosol monitoring was conducted at
one site in the Mono Lake area near Lee Vining. Although more
study is needed in this area, some preliminary remarks regarding
this data can be made. Figure 6 shows the total gravimetric mass
concentrations measured during this study period. The most

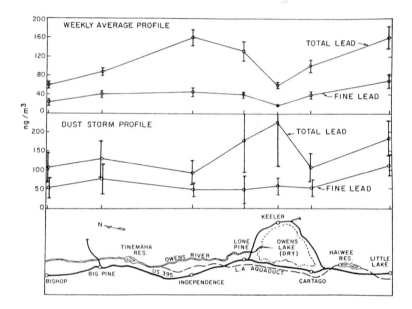

Figure 9. *Average weekly concentration at each sampling site of total (particles less than 2.5 μm) Pb.*

Error bars are standard error of the mean. Also included are profiles of total and fine Pb for the dust storm of April 13–14, 1979. Error bars are ±15% measurement system error.

obvious trend shown by these data is the increase in TGM from the beginning of the study period to the end. This is probably due to decreased precipitation and the subsequent drying of the soil near the sampling site. We believe that if monitoring had continued, even higher TSP concentrations would have been ovserved during the summer. The peaks of late April and early May correspond to dust storm periods observed in the record for Owens Valley.

The measured sulfur levels during this study period were approximately .3µg/m^3 or about .9µg/m^3 of sulfate. Soil samples collected in the Mono Lake area indicated a sulfur to iron ratio of 1.5 - 2.0. At Owens Lake, the sulfur to iron ratio is about 0.24. The data of Table IV indicate that the Mono Lake brine contains a higher proportion of sulfate anion that did Owens Lake. These data all suggest that Mono Lake has a greater potential than the Owens Lake for a sulfur aerosol problem if the Mono Lake bed is allowed to dry out. This hypothesis is supported by studies of alkaline lake bed chemistry which indicate that sulfates are found in the efflorescent crusts in lakes in Inyo County (e.g. Deep Springs Lake), and hence, can be suspended in the atmosphere. In fact, B.F. Jones of the U.S. Geological Survey asserts that at Deep Springs Lake: "although the wind frequently stirs up clouds of salt laden dust from efflorescent crusts on the west side of the playa, the well indurated crusts of the Central Lake area do not contribute much to aeolian transport," (11). Jones also indicated that Deep Springs Lake is the best chemical analog to Mono Lake.

Further study during the summer months is needed to determine if these preliminary data are representative of the typical condition in the area.

Conclusions

The data collected in the Owens Valley during this study suggest the following conclusions. First, dust storms occur frequently, and their effect on air quality in the valley is significant. This was shown by the elevated TSP values measured during the study and the frequent reports by station operators of blowing dust. The dry lake bed was found to be an important contributor to the TSP concentrations measured downwind. The contribution of lake bed aerosols to the TSP concentration at sites as far downwind as Independence was significant. Furthermore, aerosols suspended from the lake bed included significant concentrations of sulfur. Violations of sulfate standards near the lake bed often occur during dust storms. Preliminary data collected in the Mono Lake area indicate that the dry lake bed areas near Lee Vining may contribute substantial aerosol mass to the measured TSP levels. Furthermore, the potential for sulfate standards violations due to lake bed suspended aerosols at Mono Lake may be greater than the observed violations in the Owens Valley,

since the sulfur to iron ratio is an order of magnitude greater
for Mono Lake bed soil samples than for Owens Lake bed soils.
Further study is needed in the Mono Lake area in order to determine the magnitude and extent of the aerosols suspended from the
lake bed.

Literature Cited

1. Eugster, H.P. and Hardie, L.A. "Saline Lakes" chapter 8 of
    Lakes: Chemistry, Geology, Physics, Abraham Lerman ed.,
    Springer-Verlag, New York, 1978.

2. Reinking, R.F., Mathews, L.A., and St. Amand, P. "Dust Storms
    Due to Dessication of Owens Lake," International Conference
    on Environmental Sensing and Assessment, Las Vegas, Nevada
    September 14-15, 1975, IEEE publishers 1976.

3. Slinn, "Dry Deposition and Suspension of Aerosol Particles -
    A New Look at Some Old Problems" ERDA Symposium, 1976, 38, 1.

4. Cahill, T.A., Ashbaugh, L.L., Barone, J.B., Eldred, R.A.,
    Feeney, P.J., Flocchini, R.G., Goodart, C., Shadoan, D.J.,
    and Wolfe, G. "Analysis of Respirable Fractions in Atmospheric
    Particulates via Sequential Filtration" APCA Journal 1977,
    27(7) 675.

5. Whitby, R.T., Husar, R.B., Liu, B.Y.U. "The Aerosol Size Dis-
    tribution of Los Angeles Smog" J. Colloid. Interface Sci.,
    1972, 39, 211.

6. ICRP Task Group on Lung Dynamics, "Deposition and Retention
    Models for Internal Dosimetry of the Human Respiratory Tract"
    Health Physics, 1966, 12, 173.

7. Cahill, T.A., Eldred, R.A., Barone, J.B., Ashbaugh, L.L.,
    "Ambient Aerosol Sampling with Stacked Filter Units" FHWA
    Report 1978, FHWA-RD-78-178.

8. Cahill, T.A., Flocchini, R.G., Feeney, P.J., Lang, S.,
    Shadoan, D.J., Wolfe, G. "Monitoring Smog Aerosols with Ele-
    mental Analysis by Accelerator Beams" Nat. Bur. Stand. 1976,
    Special Pub. 422, 1119.

9. Bagnold, R.A. "The Physics of Blown Sand and Desert Dunes"
    Methuen, London, 1941.

10. Clark, F.W. "The data of Geochemistry" U.S.G.S. Bull., 1924,
    770, 156.

11. Jones, B.F. U.S.G.S. Prof. Paper 502-A, 1965.

RECEIVED March 10, 1981.

INDEX

Jacket design by Carol Conway.
Production by Susan Moses and Gabriele Glang

Elements typeset by Service Composition Co., Baltimore, MD.
Printed and bound by The Maple Press Co., York, PA.

R

)